Communications and Control Engineering

For other titles published in this series, go to
www.springer.com/series/61

Series Editors

A. Isidori • J.H. van Schuppen • E.D. Sontag • M. Thoma • M. Krstic

Published titles include:

Stability and Stabilization of Infinite Dimensional Systems with Applications
Zheng-Hua Luo, Bao-Zhu Guo and Omer Morgul

Nonsmooth Mechanics (Second edition)
Bernard Brogliato

Nonlinear Control Systems II
Alberto Isidori

L_2-Gain and Passivity Techniques in Nonlinear Control
Arjan van der Schaft

Control of Linear Systems with Regulation and Input Constraints
Ali Saberi, Anton A. Stoorvogel and Peddapullaiah Sannuti

Robust and H_∞ Control
Ben M. Chen

Computer Controlled Systems
Efim N. Rosenwasser and Bernhard P. Lampe

Control of Complex and Uncertain Systems
Stanislav V. Emelyanov and Sergey K. Korovin

Robust Control Design Using H_∞ Methods
Ian R. Petersen, Valery A. Ugrinovski and Andrey V. Savkin

Model Reduction for Control System Design
Goro Obinata and Brian D.O. Anderson

Control Theory for Linear Systems
Harry L. Trentelman, Anton Stoorvogel and Malo Hautus

Functional Adaptive Control
Simon G. Fabri and Visakan Kadirkamanathan

Positive 1D and 2D Systems
Tadeusz Kaczorek

Identification and Control Using Volterra Models
Francis J. Doyle III, Ronald K. Pearson and Babatunde A. Ogunnaike

Non-linear Control for Underactuated Mechanical Systems
Isabelle Fantoni and Rogelio Lozano

Robust Control (Second edition)
Jürgen Ackermann

Flow Control by Feedback
Ole Morten Aamo and Miroslav Krstic

Learning and Generalization (Second edition)
Mathukumalli Vidyasagar

Constrained Control and Estimation
Graham C. Goodwin, Maria M. Seron and José A. De Doná

Randomized Algorithms for Analysis and Control of Uncertain Systems
Roberto Tempo, Giuseppe Calafiore and Fabrizio Dabbene

Switched Linear Systems
Zhendong Sun and Shuzhi S. Ge

Subspace Methods for System Identification
Tohru Katayama

Digital Control Systems
Ioan D. Landau and Gianluca Zito

Multivariable Computer-controlled Systems
Efim N. Rosenwasser and Bernhard P. Lampe

Dissipative Systems Analysis and Control (Second edition)
Bernard Brogliato, Rogelio Lozano, Bernhard Maschke and Olav Egeland

Algebraic Methods for Nonlinear Control Systems
Giuseppe Conte, Claude H. Moog and Anna M. Perdon

Polynomial and Rational Matrices
Tadeusz Kaczorek

Simulation-based Algorithms for Markov Decision Processes
Hyeong Soo Chang, Michael C. Fu, Jiaqiao Hu and Steven I. Marcus

Iterative Learning Control
Hyo-Sung Ahn, Kevin L. Moore and YangQuan Chen

Distributed Consensus in Multi-vehicle Cooperative Control
Wei Ren and Randal W. Beard

Control of Singular Systems with Random Abrupt Changes
El-Kébir Boukas

Nonlinear and Adaptive Control with Applications
Alessandro Astolfi, Dimitrios Karagiannis and Romeo Ortega

Stabilization, Optimal and Robust Control
Aziz Belmiloudi

Control of Nonlinear Dynamical Systems
Felix L. Chernous'ko, Igor M. Ananievski and Sergey A. Reshmin

Periodic Systems
Sergio Bittanti and Patrizio Colaneri

Discontinuous Systems
Yury V. Orlov

Constructions of Strict Lyapunov Functions
Michael Malisoff and Frédéric Mazenc

Controlling Chaos
Huaguang Zhang, Derong Liu and Zhiliang Wang

Stabilization of Navier–Stokes Flows
Viorel Barbu

Distributed Control of Multi-agent Networks
Wei Ren and Yongcan Cao

He Bai • Murat Arcak • John Wen

Cooperative Control Design

A Systematic, Passivity-Based Approach

Springer

He Bai
UtopiaCompression Corporation
Los Angeles, CA
USA
he@utopiacompression.com

Murat Arcak
University of California-Berkeley
Berkeley, CA
USA
arcak@eecs.berkeley.edu

John Wen
Rensselaer Polytechnic Institute
Troy, NY
USA
wenj@rpi.edu

ISSN 0178-5354
ISBN 978-1-4614-0013-4 e-ISBN 978-1-4614-0014-1
DOI 10.1007/978-1-4614-0014-1
Springer New York Dordrecht Heidelberg London

British Library Cataloguing in Publication Data
A catalogue record for this book is available from the British Library

Library of Congress Control Number: 2011929229

Mathematics Subject Classification (2010): 93-02, 92C10, 93D15, 49M37

Cover design: VTeX UAB, Lithuania

Printed on acid-free paper

Springer is part of Springer Science+Business Media (www.springer.com)

To Xiaoqing
H.B.

To Stephanie
M.A.

To Debbie
J.T.W.

Preface

Recent advances in sensing, communication and computation technologies have enabled a group of agents, such as robots, to communicate or sense their relative information and to perform tasks in a collaborative fashion. The past few years witnessed rapidly-growing research in cooperative control technology. Applications range from space interferometry sensing to environmental monitoring, to distributed computing, and distributed target surveillance and tracking. However, the analytical techniques used in cooperative control algorithms have been disparate, and a unified theory has been wanting.

In this book, we present a passivity-based framework that allows a systematic design of scalable and decentralized cooperative control laws. This framework makes explicit the passivity properties used implicitly in existing results and simplifies the design and analysis of a complex network of agents by exploiting the network structure and inherent passivity properties of agent dynamics. As we demonstrate in the book, this passivity-based framework can be easily tailored to handle classes of cooperative control problems. Each of these problems has important applications in practice. For example, formation control and coordinated path following allow vehicles to maintain a tight relative configuration, thereby achieving effective group sensing or drag force reduction. Attitude coordination ensures that multiple spacecraft keep a precise relative attitude, which is important in space interferometry missions. Cooperative load transport enables robots to move an object around meanwhile exerting desired internal forces on the object.

To demonstrate the design flexibility offered by the passivity-based framework, we feature various adaptive redesigns. Adaptivity is important in multi-agent systems, as it implies that the agents can adjust their control according to changing conditions. A particularly useful scenario is when a group leader has the group mission plan while the remaining agents have only partial information of the plan. As the leader autonomously modifies the mission plan, it is essential that the rest of the group be able to adapt. We develop adaptive designs with which the agents recover the leader's mission plan. Another scenario that requires adaptivity is when uncertainty exists in agents' controls, such as wind and viscous damping. In this case, the agents should adapt their control laws to simultaneously accommodate this uncertainty and achieve the required task. We illustrate this scenario with adaptive designs for agreement of multiple Euler-Lagrange systems.

The intention of this book is to summarize in a coherent manner the authors' recent research in passivity-based cooperative control of multi-agent systems. Related passivity approaches are applicable to other interconnected systems, such as data communication networks, biomolecular systems, building heating, ventilation, and air conditioning (HVAC) systems, and power networks, though these applications are outside the scope of this book. The organization of this book is as follows:
Chapter 1 introduces cooperative control and presents the necessary background in graph theory and passivity.
Chapter 2 presents the passivity-based approach to cooperative control and applies it to the agreement problem. We illustrate a position-based formation control problem that can be transformed into the agreement problem. We also study a distance-based formation control under the same passivity-based framework and compare it with the position-based formation control.
Chapters 3 and 4 consider the situation where only one agent possesses the group reference velocity information and develop adaptive designs that recover such information for the other agents. Chapter 3 adopts an internal model approach while Chapter 4 parameterizes the reference velocity with time-varying basis functions. These two adaptive schemes illustrate the design flexibility offered by the passivity-based approach.
Chapter 5 investigates attitude agreement of multiple rigid bodies. We adapt the results in Chapters 2 and 4 to a similar passivity-based framework that addresses the agreement of agents in $SO(3)$.
Chapter 6 studies agreement problem of multiple Euler-Lagrange systems. In particular, we consider the case where uncertainty exists in agents' dynamics. We present two adaptive designs that ensure the agreement in the presence of uncertainty.
Chapter 7 presents a synchronized path following problem, where the agents achieve tracking of a desired formation by synchronizing path parameters. The synchronization of the path parameters is solved by the passivity-based agreement designs in Chapter 2.
Chapter 8 studies cooperatively transporting a common load by multiple robots. We formulate this problem in a similar fashion to the position-based formation control in Chapter 2 and validate it with experiments.
Chapter 9 provides an investigation of robustness properties for a second order linear cooperative control system.
The Appendix includes technical proofs and tools utilized in this book.

The results in this book would not have been possible without the help of our colleagues and collaborators. We are indebted to Professors Petar Kokotović, Panagiotis Tsiotras, Derek Paley, Ming Cao, Arthur Sanderson, and A. Agung Julius for insightful comments on parts of this work. We also acknowledge Emrah Biyik, Ivar-André Ihle and Thor Fossen, whose joint work with the second author are presented in Section 4.5 and in Chapter 7. We would like to express our gratitude to Mehran Mesbahi, Wei Ren, Fumin Zhang and Sam Coogan for reviewing parts of the manuscript and for providing valuable feedback. We also thank our editors Steven Elliot and Andrew Leigh for their interest and efforts in publishing this book. The first author is grateful to Professors Kevin Lynch and Randy Freeman for offering him a precious opportunity to continue working on cooperative control as a research fellow at Northwestern University and for their constant guidance and support. He also thanks John Wason and Karen Chapin for help in experimental implementation, and Aranya Chakrabortty, Fabio Morbidi, Florian Dörfler, Chinpei Tang, and Tiejun Zhang for helpful discussions during the preparation of the book.

We acknowledge the financial support by the Air Force Office of Scientific Research under grants FA9550-09-1-0092 and FA9550-07-1-0308 and by the National Science Foundation under grant ECCS-0852750. This work was also supported partly by the Center for Automation Technologies and Systems (CATS) at Rensselaer Polytechnic Institute under a block grant from the New York State Office of Science, Technology, and Academic Research (NYSTAR).

He Bai UtopiaCompression Corporation, Los Angeles, California
Murat Arcak University of California, Berkeley, California
John T. Wen Rensselaer Polytechnic Institute, Troy, New York

Contents

Chapter 1
Introduction

1.1 What Is Cooperative Control?

Multiple robots or sensors have a number of advantages over a single agent, including robustness to failures of individual agents, reconfigurability, and the ability to perform challenging tasks such as environmental monitoring, target localization, that cannot be achieved by a single agent. A cooperative control system consists of a group of autonomous agents with sensing or communication capabilities, for example, robots with camera sensors and vehicles with communication devices. The goal of the group is to achieve prescribed agent and group behaviors using only local information available to each agent from sensing or communication devices. Such local information may include relative configuration and motion obtained from sensing or communication between agents, agent's sensor measurements, and so on. The relative sensing and communication dictates the architecture of information flow between agents. Thus, a cooperative system has four basic elements: group objective, agents, information topology, and control algorithms governing the motion of the agents.

Examples of group objectives in cooperative control include flocking, schooling, cohesion, guarding, escorting [112, 98, 131, 76, 51, 107], agreement [110, 75, 102, 63, 10, 103, 5], vehicle formation maintenance [124, 47, 36, 5], gradient climbing in an environmental field [8, 98, 27, 18], cooperative load transport [14, 145, 106, 129, 128], distributed estimation and optimal sensing [97, 99, 86, 153, 33], source seeking [151] and coverage control [34, 78]. Some of the cooperative control objectives involve only relative variables (e.g., relative positions) between agents while others depend on absolute variables (e.g., inertial positions) of agents. We illustrate below some examples of cooperative control.

Formation Control. A major focus in cooperative control is formation stability, where the group objective is to stabilize the relative distances/positions between the agents to prescribed desired values. Formation maintenance finds natural applications in coverage control, drag force reduction, and space interferometry. We take space interferometry as a motivating example. As shown in Fig. 1.1, space in-

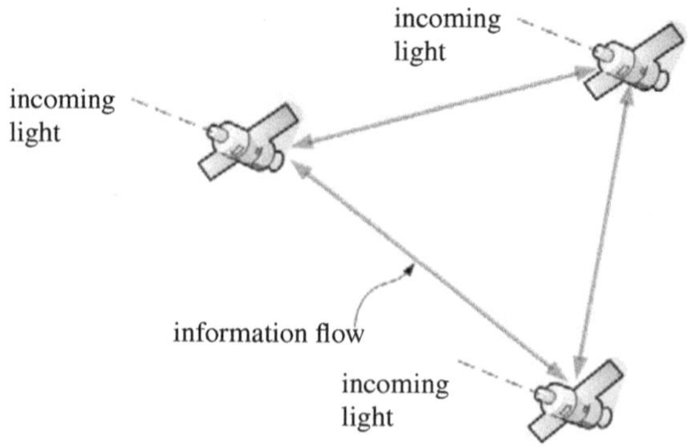

Fig. 1.1 Space interferometry using multiple spacecraft. Spacecraft must maintain precise relative formation and the same attitude towards the incoming light to generate an interferometry pattern. The information flow between spacecraft is setup by optical sensors, which measure relative positions between spacecraft.

terferometry uses multiple spacecraft pointing towards an object of interest. Each spacecraft collects light from the object. If the relative positions between spacecraft are maintained precisely, an interferometry pattern is generated and measurement of the magnitude and phase of this pattern can be obtained by coherently mixing the light from individual spacecraft [73, 118, 123]. Multiple such measurements allow reconstruction of the image of the object, which has a much finer resolution than any single spaceborne telescope achieves. In space, position information of individual spacecraft in the earth frame is imprecise or unavailable, whereas relative position between spacecraft can be measured precisely by optical sensors [74]. Thus, maintaining spacecraft formation in space may make use of only relative position information while maneuvering the formation to point towards objects of interest.

Agreement. In the agreement problem, the group objective is the convergence of distributed variables of interest (agents' positions or headings, phase of oscillations, etc.) to a common value. Given different contexts, the agreement problem is also called *consensus*, or *synchronization*. As shown in Fig. 1.1, space interferometry requires spacecraft to align their attitudes with each other, which is an agreement problem. Agreement problem also has potential applications in schooling and flocking in distributed robotics and biological systems [100, 112, 127, 24], distributed estimation and sensor fusion [99, 104], fire surveillance [25] and distributed computing [148, 140, 16], among others.

Optimal Sensing. The group objective for optimal sensing is to optimally place the agents' positions so that certain meaningful utility functions are maximized. Examples of utility functions include probability of detecting a target [34, 78] and information obtained from a sensor network [86, 33]. In this case, the utility functions usually depend on the absolute positions of the agents.

Most cooperative control problems concern coordinated motion of agents in different scenarios. Therefore, agent dynamics become important in achieving different group objectives. Small mobile sensors or robots can be controlled by directly manipulating their velocities. Such agents are commonly modeled by a first order kinematic model. Depending on the group objective, the agent model can be a simple integrator or a set of integrators subject to nonholonomic constraints, such as a unicycle model. If the group objective is to control the position of the sensor to improve sensing capability, it may suffice to model the sensors as massless points with single integrator kinematics. When the velocities of agents are not directly manipulatable or the masses of the agents are significant, double integrator agent dynamics are more appropriate. In numerous applications, the attitudes of the agents play an important role, which means that agent models in Euler-Lagrangian form or in Hamiltonian form may be required.

To achieve the group objective, each agent may need information from other agents. If agent i has access to agent j's information, the information of agent j flows to agent i and agent j is a neighbor of agent i. The abstract structure of the information flows in the group is then represented as a graph, where each agent is a node and the information flows between them are represented as links [1]. In many applications, the information flow between agents is achieved by direct sensing or communication. For example, to control the relative distances between the agents, the agents obtain their relative distance information by sensing or by communicating their inertial positions. In some applications, the information flow is realized through a physical medium. For example, consider a group of agents transporting a payload. By interacting with the payload, the agents can obtain the relative information between them without explicit communication. We will illustrate such an example in Chapter 8.

One of the main challenges in cooperative control design is to achieve prescribed group objectives by *distributed* feedback laws. The distributed laws make use of information available only to individual agents. Such information includes the information flow from neighbors, and sensor measurements from agent itself. Take the agreement problem as an example. When modeled as a first order integrator, the agent can aggregate the differences between its own state and its neighbors' and take that aggregated value as a feedback control. In this case, the control algorithm is distributed since it only employs information from neighboring agents. If the control algorithm and the agent model are both linear, stability can be analyzed by examining the eigenvalues of the closed-loop system matrix with the help of algebraic graph theory. This approach leads to simple stability criteria for the agreement problem, e.g., [63, 111, 102, 103, 47, 76, 87].

However, for some applications of cooperative control, only nonlinear algorithms can achieve the objective. Consider the following formation control problem: The group objective is to stabilize relative distances (the Euclidean norms of relative positions) between agents to desired values. In this case, the desired equilibria are spheres, which are compact sets containing more than one point. When each agent is modeled as a linear system, such as a double integrator, there is no linear feedback law globally stabilizing the desired equilibria. This is because a linear agent model with linear feedback results in a linear system, whose equilibria can simply

[1] A brief introduction to graph theory will be presented later in this chapter.

be a point or a subspace. Thus, only nonlinear feedback laws may solve this formation control problem. Indeed, most of the formation control algorithms have been proposed in the form of nonlinear artificial attraction and repulsion forces between neighboring agents. The design and analysis of such rules make use of graph theory and potential function methods.

1.2 What Is in This Book?

For different cooperative control problems, there are different control design methods. In this book, we introduce a *unifying* passivity-based framework for cooperative control problems. Under this passivity-based framework, we develop robust, adaptive, and scalable design techniques that address a broad class of cooperative control problems, including the formation control and the agreement problem discussed above.

This framework makes explicit the passivity properties used implicitly in the Lyapunov analysis of several earlier results, including [131, 98, 102], and simplifies the design and analysis of a complex network of agents by exploiting the network structure and inherent passivity properties of agent dynamics. With this simplification, the passivity approach further overcomes the simplifying assumptions of existing designs and offers numerous advantages, including:

1. Admissibility of complex and heterogenous agent dynamics: Unlike some of the existing cooperative control literature where the agent is modeled as a point robot, the passivity approach allows high order and nonlinear dynamics, including Lagrangian and Hamiltonian systems. As illustrated in Chapter 5, attitude coordination among multiple rigid bodies can be studied under this passivity framework. In this case, the agent dynamics are in the Hamiltonian form. Chapter 6 discusses the agreement of multiple Lagrangian systems. The passivity approach is further applicable to *heterogenous* systems in which the agent dynamics and parameters, such as masses, dampings, vary across the group.

2. Design flexibility, robustness and adaptivity: The passivity approach abstracts the common core of several multi-agent coordination problems, such as formation stabilization, group agreement, and attitude coordination. Because passivity involves only input-output variables, it has inherent robustness to unknown model parameters. Since passivity is closely related to Lyapunov stability, this passivity approach lends itself to systematic adaptive designs that enhance robustness of cooperative systems. Such design flexibility and adaptivity will be demonstrated in this book by the adaptive designs in Chapters 3, 4 and 6.

3. Modularity and scalability: The passivity framework yields decentralized controllers which allow the agents to make decisions based on relative information with respect to their neighbors, such as relative distance. A key advantage of the passivity-based design is its *modularity*, which means that the control laws do not rely on the knowledge of number of other agents, the communication structure of the network, or any other global network parameters.

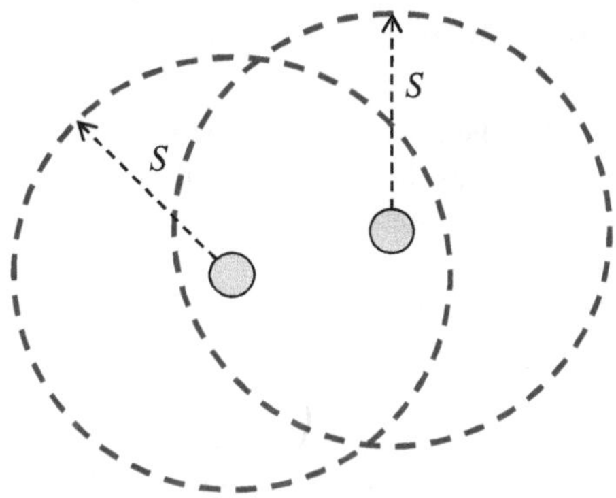

Fig. 1.2 If the sensing/communication ranges for both robots are chosen to be the same and one agent is within the sensing/communication range of the other agent, the information flow between them is symmetric. $S > 0$ denotes the sensing or communication radius.

Our major assumptions for this passivity-based approach are:

Bidirectional Information Topology: Control algorithms with bidirectional information topology tend to have inherent stability properties as we explicate with the help of passivity arguments in this book. Although directional information topology can render stability for first order agents [103, 109], it may lead to instability for high order agents, as we illustrate in Example 2.3 in Chapter 2.

Bidirectional information topology also appears naturally in a number of cooperative control applications. For example, as shown in Fig. 1.2, the information topology of agents with the same sensing range can be modeled as bidirectional. In the load transport problem studied in Chapter 8, the agents exert force on the payload and receive reaction forces from the payload. The exerted force and the reaction force contain implicitly the relative motion information between the agents and the payload. Thus, the information flows are bidirectional.

Static Information Topology: Our design assumes that the information topology remains unchanged. This is not a restrictive assumption since in most practical situations, the information topology remains static for a certain period of time. If that period of time is long enough, by standard dwell time arguments [91, 56], the closed-loop system remains stable. Note that for first order linear consensus protocols, such as those studied in [103, 63], robustness to arbitrary switching topology has been justified. However, for higher order systems, it is well known that switching may lead to instability [80, 81]. We will show that for first order protocols, the passivity-based framework can handle a broad class of switching topology whereas for higher order cooperative systems, topology switching improperly may result in instability.

1.3 Notation and Definition

• We denote by $\mathbb{R}$ and by $\mathbb{C}$ the set of real numbers and complex numbers, respectively. The notation $\mathbb{R}_{\geq 0}$ denotes the set of all real nonnegative numbers. The real part and the imaginary part of a complex number $x \in \mathbb{C}$ are given by $Re[x]$ and $Im[x]$, respectively.
• All the vectors in this book are column vectors. The set of p by 1 real vectors is denoted by $\mathbb{R}^p$ while the set of p by q real matrices is denoted by $\mathbb{R}^{p\times q}$. The transpose of a matrix $A \in \mathbb{R}^{p\times q}$ is given by $A^T \in \mathbb{R}^{q\times p}$.
•$\mathcal{N}(A)$ and $\mathcal{R}(A)$ are the null space (kernel) and the range space of a matrix A, respectively. I_p and $\mathbf{0}_p$ denote the $p\times p$ identity and zero matrices, respectively. The $p\times q$ zero matrix is denoted by $\mathbf{0}_{p\times q}$. Likewise, 1_N and 0_N denote the $N\times 1$ vector with each entry of 1 and 0, respectively. Without confusion, we will also use 0 to denote a vector of zeros with a compatible dimension.
• The Kronecker product of matrices $A \in \mathbb{R}^{m\times n}$ and $B \in \mathbb{R}^{p\times q}$ is defined as

$$A \otimes B := \begin{bmatrix} a_{11}B & \cdots & a_{1n}B \\ \vdots & \ddots & \vdots \\ a_{m1}B & \cdots & a_{mn}B \end{bmatrix} \in \mathbb{R}^{mp\times nq}, \tag{1.1}$$

and satisfies the properties

$$(A\otimes B)^T = A^T \otimes B^T \tag{1.2}$$

$$(A\otimes I_p)(C\otimes I_p) = (AC)\otimes I_p \tag{1.3}$$

where A and C are assumed to be compatible for multiplication.
• The maximum and minimum eigenvalues of a symmetric matrix A are denoted by $\lambda_{\max}(A)$ and $\lambda_{\min}(A)$, respectively.
• For a vector $x \in \mathbb{R}^p$, $|x|$ denotes its 2-norm, that is $|x| = \sqrt{x^T x}$.
• The norm of a matrix A is defined as its induced norm $||A|| = \sqrt{\lambda_{\max}(A^T A)}$.
• We use the notation $\mathrm{diag}\{K_1, K_2, \cdots, K_n\}$ to denote the block diagonal matrix

$$\begin{pmatrix} K_1 & \mathbf{0}_{p\times q} & \cdots & \mathbf{0}_{p\times q} \\ \mathbf{0}_{p\times q} & K_2 & \cdots & \mathbf{0}_{p\times q} \\ \vdots & \vdots & \ddots & \vdots \\ \mathbf{0}_{p\times q} & \mathbf{0}_{p\times q} & \cdots & K_n \end{pmatrix} \tag{1.4}$$

where $K_i \in \mathbb{R}^{p\times q}$, $i = 1, \cdots, n$.
• The notation $K = K^T > 0$ means that K is a symmetric positive definite matrix while $k > 0$ implies k is a positive scalar.
• Given a vector $v \in \mathbb{R}^3$, the cross product $v\times$ is a linear operator, and can be represented in a coordinate frame as left-multiplication by the skew-symmetric matrix:

$$\widehat{v} = \begin{bmatrix} 0 & -v_3 & v_2 \\ v_3 & 0 & -v_1 \\ -v_2 & v_1 & 0 \end{bmatrix} \tag{1.5}$$

where (v_1, v_2, v_3) are the components of v in the given coordinate frame. The inverse operation of cross product is given by $^\vee$, that is

$$(\widehat{v})^\vee = v. \tag{1.6}$$

• For the coordinate frame representation of a vector, the leading superscript indicates the reference frame while the subscript i denotes the agent i. The superscript d means the desired value. As an illustration, ${}^j v_i^d$ means the desired velocity of the ith agent in the jth frame.
• A function is said to be C^k if its partial derivatives exist and are continuous up to order k.
• Given a C^2 function $P : \mathbb{R}^p \rightarrow \mathbb{R}$ we denote by ∇P its gradient vector, and by $\nabla^2 P$ its Hessian matrix.
• A function $\alpha : [0, a) \rightarrow \mathbb{R}_{\geq 0}$ is of class K if it is continuous, strictly increasing and satisfies $\alpha(0) = 0$. It is said to belong to class K_∞ if $a = \infty$ and $\alpha(r) \rightarrow \infty$ as $r \rightarrow \infty$. A function $\beta : \mathbb{R}_{\geq 0} \times \mathbb{R}_{\geq 0} \rightarrow \mathbb{R}_{\geq 0}$ is of class KL if, for each fixed s, the function $\beta(r,s)$ belongs to class K with respect to r and, for each fixed r, the function $\beta(r,s)$ is decreasing with respect to s and $\beta(r,s) \rightarrow 0$ as $s \rightarrow \infty$. An example of class KL functions is shown in Fig. 1.3.

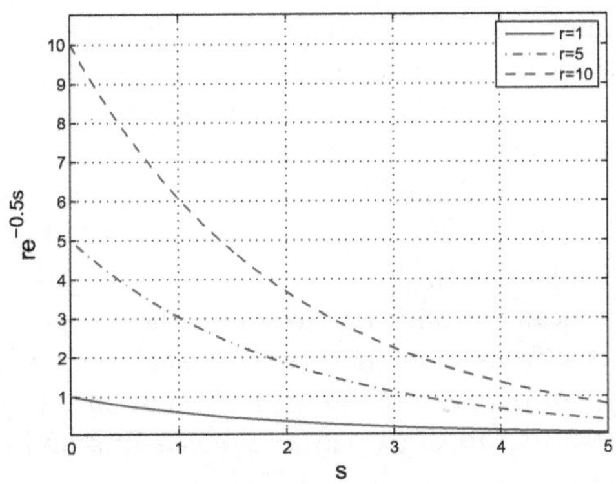

Fig. 1.3 The function $\beta(r,s) = re^{-0.5s}$ is of class KL because for fixed r, $re^{-0.5s}$ is decreasing and converges to zero as s converges to ∞ and for fixed s, $re^{-0.5s}$ is monotonically increasing with respect to r.

• The system $\dot{x} = f(x,u)$ is said to be Input-to-State Stable (ISS) [125, 126] if there exist functions $\beta \in KL$, $\rho \in K$ such that for any initial state $x(t_0)$ and any bounded input $u(t)$, the solution $x(t)$ exists for all $t \geq 0$ and satisfies

$$|x(t)| \leq \beta(|x(t_0)|, t - t_0) + \rho(\sup_{t_0 \leq \tau \leq t} |u(\tau)|). \tag{1.7}$$

• For a closed set $\mathscr{A}$, $|\chi|_{\mathscr{A}}$ denotes the distance from the point χ to $\mathscr{A}$, defined as

$$|\chi|_{\mathscr{A}} = \inf_{\eta \in \mathscr{A}} |\chi - \eta|. \tag{1.8}$$

• Given the dynamics of the state $\chi(t)$, a closed invariant set $\mathscr{A}$ is uniformly asymptotically stable with region of attraction $\mathscr{G}$ if for each $\varepsilon > 0$ there exists $\delta > 0$ such that

$$|\chi(t_0)|_{\mathscr{A}} \leq \delta \quad \Rightarrow \quad |\chi(t)|_{\mathscr{A}} \leq \varepsilon \quad \forall t \geq t_0 \tag{1.9}$$

and, if for each $\varepsilon > 0$ and $r > 0$, there exists $T > 0$ such that for every initial condition $\chi(t_0) \in \mathscr{G}$ the resulting trajectory satisfies

$$|\chi(t_0)|_{\mathscr{A}} \leq r \Rightarrow \quad |\chi(t)|_{\mathscr{A}} \leq \varepsilon \quad \forall t \geq T. \tag{1.10}$$

Several results on set stability and, in particular, converse Lyapunov theorems are presented in [82] and [137].
• We use the notation $\chi \to \partial\mathscr{G}^{\infty}$ to indicate a sequence of points χ in $\mathscr{G}$ converging to a point on the boundary of $\mathscr{G}$, or if $\mathscr{G}$ is unbounded, having the property $|\chi| \to \infty$.

1.4 Basic Graph Theory

In this book, we will make use of basic result from algebraic graph theory to facilitate our analysis. The results presented in this section are standard in the literature and will be well known to readers familiar with graph theory.

A graph is an abstract representation of a group of nodes where some of them are connected by links. More formally, a graph G is an ordered pair $G = (\mathscr{V}, E)$ consisting of a set $\mathscr{V}$ of nodes and a set $E \subset \mathscr{V} \times \mathscr{V}$ of links. Thus, a link is an ordered pair of two *distinct* nodes.

A directed link (i, j) is an *incoming link* to node j and an *outgoing link* from node i. We then draw an arrow from node i to node j. We call node i (respectively, j) the *negative* (respectively, *positive*) end of link (i, j). If both links (i, j) and (j, i) belong to E, we combine these two links as one *undirected* link and use a bidirectional arrow to denote this link.

Depending on the directions of the links, a graph may be categorized as *directed* or *undirected*. If a graph G consists of only undirected links, it is *undirected*. Otherwise, the graph is *directed*.

We say node i is a *neighbor* of node j if the link (i, j) exists in the graph G. This means that for each directional link, the negative end is the neighbor of the positive end. Note that for undirected graphs, if node i is a neighbor of node j, then node j is also a neighbor of node i. We denote by $\mathscr{N}_j$ the *set of neighbors* of node j.

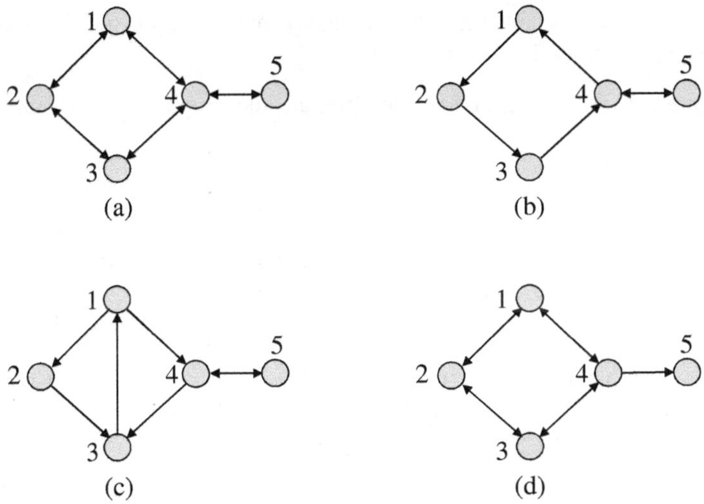

Fig. 1.4 Different types of graphs of five nodes. (a): an undirected connected graph. (b): a balanced and strongly connected graph. (c): a strongly connected graph. (d): a weakly connected graph. A directed link is denoted by a line with a directional arrow while an undirected link is denoted by a bidirectional arrow. The node number is beside each node.

For $i \in \mathscr{V}$, if the number of incoming links to i is the same as the number of outgoing links from i, the graph is *balanced*. Clearly, an undirected graph is a special balanced graph.

A *directed path* is a sequence of p nodes $1, \cdots, p$, such that $(i, i+1) \in E$, $\forall i = 1, \cdots, p-1$. A *cycle* is a directed path such that the starting and the ending nodes of the path are the same. A graph is called *strongly connected* if there exists a directed path from any one node to another. Note that for an undirected graph, strong connectedness is simply termed connectedness. A graph is called *weakly connected* if replacing all the directed links in E with undirected ones gives a connected undirected graph. In Fig. 1.4 are several examples of five nodes illustrating connectedness of different graphs.

Definition 1.1 (Graph Laplacian matrix L).
Consider a directed graph G with N nodes. The Laplacian matrix of a graph G, denoted by $L \in \mathbb{R}^{N \times N}$, is given by

$$\ell_{ij} := \begin{cases} |\mathscr{N}_i| & \text{if } i = j \\ -1 & \text{if } j \in \mathscr{N}_i \\ 0 & \text{otherwise,} \end{cases} \tag{1.11}$$

where $|\mathscr{N}_i|$ is the cardinality of the set $\mathscr{N}_i$. □

The definition in (1.11) results in the following property of L:

Property 1.1. The graph Laplacian matrix L has an eigenvalue of zero associated with an eigenvector 1_N, i.e., $L1_N = 0_N$. □

Example 1.1. Following (1.11), we compute the graph Laplacian matrices for the graphs in Fig. 1.4 as

$$L_a = \begin{pmatrix} 2 & -1 & 0 & -1 & 0 \\ -1 & 2 & -1 & 0 & 0 \\ 0 & -1 & 2 & -1 & 0 \\ -1 & 0 & -1 & 3 & -1 \\ 0 & 0 & 0 & -1 & 1 \end{pmatrix}, \quad L_b = \begin{pmatrix} 1 & 0 & 0 & -1 & 0 \\ -1 & 1 & 0 & 0 & 0 \\ 0 & -1 & 1 & 0 & 0 \\ 0 & 0 & -1 & 2 & -1 \\ 0 & 0 & 0 & -1 & 1 \end{pmatrix} \tag{1.12}$$

$$L_c = \begin{pmatrix} 1 & 0 & -1 & 0 & 0 \\ -1 & 1 & 0 & 0 & 0 \\ 0 & -1 & 2 & -1 & 0 \\ -1 & 0 & 0 & 2 & -1 \\ 0 & 0 & 0 & -1 & 1 \end{pmatrix}, \quad \text{and} \quad L_d = \begin{pmatrix} 2 & -1 & 0 & -1 & 0 \\ -1 & 2 & -1 & 0 & 0 \\ 0 & -1 & 2 & -1 & 0 \\ -1 & 0 & -1 & 2 & 0 \\ 0 & 0 & 0 & -1 & 1 \end{pmatrix}. \tag{1.13}$$

It is easy to see all these four Laplacian matrices satisfy Property 1.1. □

In particular, the Laplacian matrix for undirected graphs satisfies Properties 1.2 and 1.3 below.

Property 1.2. The Laplacian matrix L of an undirected graph is symmetric and positive semidefinite. □

Property 1.3. [17, Item 4e and Corollary 6.5]
An undirected graph is connected if and only if the second smallest eigenvalue of its Laplacian matrix is strictly positive. □

We verify the positive semidefiniteness of L in Property 1.2 by showing

$$y^T Ly \geq 0 \quad \forall y \in \mathbb{R}^N. \tag{1.14}$$

To see this, we let y_i be the ith element of y and note from (1.11) that

$$y^T Ly = \sum_{i=1}^{N} y_i \sum_{j \in \mathcal{N}_i} (y_i - y_j) = \sum_{i=1}^{N} \sum_{j \in \mathcal{N}_i} (y_i^2 - y_i y_j) \tag{1.15}$$

$$= \sum_{i=1}^{N} \sum_{j \in \mathcal{N}_i} (y_i^2 - 2y_i y_j + y_j^2) + \sum_{i=1}^{N} \sum_{j \in \mathcal{N}_i} (y_i y_j - y_j^2) \tag{1.16}$$

$$= \sum_{i=1}^{N} \sum_{j \in \mathcal{N}_i} (y_i - y_j)^2 + \sum_{i=1}^{N} \sum_{j \in \mathcal{N}_i} (y_i y_j - y_j^2). \tag{1.17}$$

Because the graph is undirected, we have

$$\sum_{i=1}^{N} \sum_{j \in \mathcal{N}_i} y_j^2 = \sum_{i=1}^{N} \sum_{j \in \mathcal{N}_i} y_i^2 \tag{1.18}$$

which implies that the last term in (1.17) is indeed $-y^T Ly$. Therefore, it follows from (1.17) that

$$y^T Ly = \frac{1}{2}\sum_{i=1}^{N}\sum_{j\in\mathcal{N}_i}(y_i - y_j)^2 \geq 0. \tag{1.19}$$

For a general directed graph, the graph Laplacian matrix L is not symmetric and $y^T Ly$ can be sign-indefinite. However, if the directed graph is balanced and strongly connected, $y^T Ly \geq 0$ holds for any y due to the following property:

Property 1.4. [103] The graph Laplacian matrix L of a *balanced and strongly connected* graph G satisfies

$$L + L^T = \frac{1}{2}L_{\mathrm{sym}} \tag{1.20}$$

where L_{sym} represents the graph Laplacian matrix of the *undirected* graph obtained by replacing the directed edges in G with undirected ones. □

For an undirected graph G, we may assign an orientation to G by considering one of the two nodes of a link to be the positive end. We denote by $\mathcal{L}_i^+$ ($\mathcal{L}_i^-$) the set of links for which node i is the positive (negative) end.

Definition 1.2 (Graph Incidence matrix D).

Denoting by ℓ the total number of links, we define the $N \times \ell$ *incidence matrix* D of an undirected graph G as

$$d_{ik} := \begin{cases} +1 \text{ if } k \in \mathcal{L}_i^+ \\ -1 \text{ if } k \in \mathcal{L}_i^- \\ 0 \ \text{ otherwise.} \end{cases} \tag{1.21}$$

□

Property 1.5. We obtain from (1.21) an incidence matrix D corresponding to a particular orientation assignment to the undirected graph G. Independently of how we assign the orientation to G, the resulting incidence matrix D has the following properties:

1. The rank of D is at most $N-1$ and the rank of D is $N-1$ if and only if the graph G is connected;
2. The columns of D are linearly independent when no cycles exist in the graph;
3. If the graph G is connected, the only null space of D^T is spanned by 1_N;
4. The graph Laplacian matrix L of G satisfies

$$L = DD^T. \tag{1.22}$$

□

Example 1.2. We verify the last item in Property 1.5 by considering the graph G in Fig. 1.5. We obtain from (1.11) that

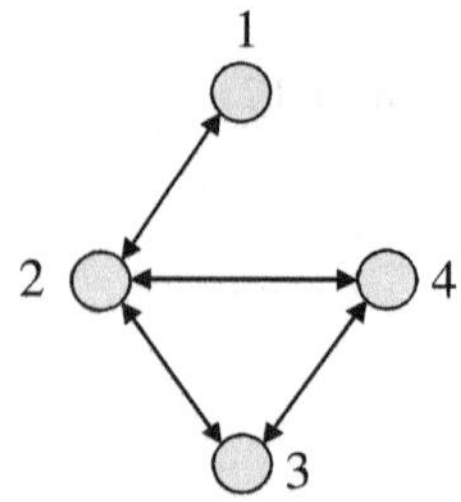

Fig. 1.5 An undirected graph of four agents whose Laplacian matrix is in (1.23). The agent number is illustrated at each node.

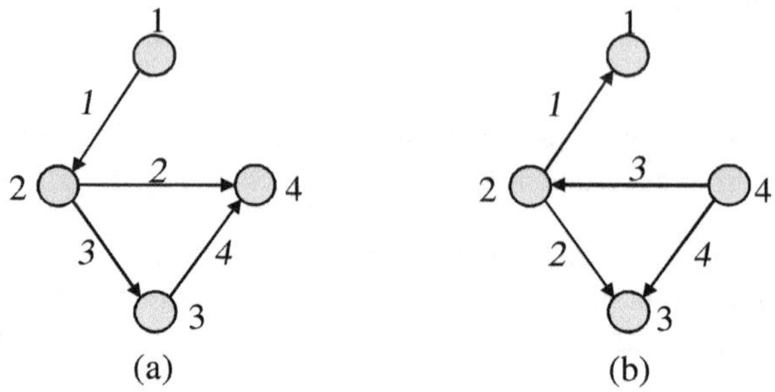

Fig. 1.6 Two different orientation assignments to the graph in Fig. 1.5 yields two different graph incidence matrices D in (1.24). However, both incidence matrices give the same Laplacian matrix (1.23) using (1.22). The arrow points to the positive end of each link. The link number is denoted in *italic* at each link.

$$L = \begin{pmatrix} 1 & -1 & 0 & 0 \\ -1 & 3 & -1 & -1 \\ 0 & -1 & 2 & -1 \\ 0 & -1 & -1 & 2 \end{pmatrix}. \tag{1.23}$$

To show that the choice of D does not affect L, we assign different orientations to G as in Fig. 1.6 and obtain the two incidence matrices D as

$$D_a = \begin{pmatrix} -1 & 0 & 0 & 0 \\ 1 & -1 & -1 & 0 \\ 0 & 0 & 1 & -1 \\ 0 & 1 & 0 & 1 \end{pmatrix} \quad \text{and} \quad D_b = \begin{pmatrix} 1 & 0 & 0 & 0 \\ -1 & -1 & 1 & 0 \\ 0 & 1 & 0 & 1 \\ 0 & 0 & -1 & -1 \end{pmatrix}. \tag{1.24}$$

A simple computation shows that $L = D_a D_a^T = D_b D_b^T$. Thus, the choice of orientation assignment to the graph does not affect the graph Laplacian matrix. □

1.5 Passivity and Passivity-preserving Structures

In this section, we briefly review the definition of passivity and its relation to stability. We also present four passivity-preserving structures that will be utilized in the rest of the book. Some of the results in this section are based on [69, 116].

Definition 1.3 (Passivity of Static Nonlinearity).
A static nonlinearity $y = h(u)$, where $h : \mathbb{R}^p \to \mathbb{R}^p$, is *passive* if, for all $u \in \mathbb{R}^p$,

$$u^T y = u^T h(u) \geq 0; \tag{1.25}$$

and *strictly passive* if (1.25) holds with strict inequality $\forall u \neq 0$. □

Definition 1.4 (Passivity and Strict Passivity of Dynamical Systems).
The dynamical system

$$\mathcal{H}: \quad \begin{cases} \dot{\xi} = f(\xi, u) \\ y = h(\xi, u), \quad \xi \in \mathbb{R}^n, \; u, y \in \mathbb{R}^p, \end{cases} \tag{1.26}$$

is said to be *passive* if there exists a C^1 *storage function* $S(\xi) \geq 0$ such that

$$\dot{S} = \nabla S(\xi)^T f(\xi, u) \leq -W(\xi) + u^T y \tag{1.27}$$

for some positive semidefinite function $W(\xi)$. We say that (1.26) is *strictly passive* if $W(\xi)$ is positive definite. □

Definition 1.5 (Strict Input and Output Passivity).
For the dynamic system (1.26), if S in (1.27) satisfies

$$\dot{S} \leq -u^T \psi(u) + u^T y \tag{1.28}$$

for some function $\psi(u)$ such that $u^T \psi(u) > 0$, then (1.26) is *input strictly passive.* Likewise, if

$$\dot{S} \leq -y^T \psi(y) + u^T y \tag{1.29}$$

holds for some function $\psi(y)$ where $y^T \psi(y) > 0$, (1.26) is *output strictly passive.* □

Example 1.3 (Passivity of Euler-Lagrange Systems).
A standard model of mechanical systems with n degrees of freedom is given by the *Euler-Lagrange equation*:

$$\frac{d}{dt}\left(\frac{\partial L}{\partial \dot{x}}(x, \dot{x})\right) - \frac{\partial L}{\partial x}(x, \dot{x}) = \tau \tag{1.30}$$

where $x = [x_1, \cdots, x_n]^T$ are the generalized coordinates of the system and $\tau = [\tau_1, \cdots, \tau_n]^T$ is the generalized torque acting on the system. The Lagrangian function $L(x, \dot{x})$ satisfies

$$L(x,\dot{x}) = K(x,\dot{x}) - P(x) \tag{1.31}$$

where $P(x)$ is the potential energy of the system and is bounded from below, i.e.,

$$P(x) \geq \bar{P} := \min_x P(x), \tag{1.32}$$

and $K(x,\dot{x})$ is the kinetic energy of the system which is assumed to be of the form

$$K(x,\dot{x}) = \frac{1}{2}\dot{x}^T M(x)\dot{x}, \tag{1.33}$$

in which $M(x) = M(x)^T$ is the positive definite generalized inertia matrix.
A further computation from (1.30) and (1.33) leads to

$$M(x)\ddot{x} + C(x,\dot{x})\dot{x} + g(x) = \tau \tag{1.34}$$

where $g(x) = \frac{\partial P(x)}{\partial x}$. A well known property of (1.34) is that $\dot{M}(x) - 2C(x,\dot{x})$ is skew-symmetric [6, 71], i.e.,

$$\dot{M}(x) - 2C(x,\dot{x}) = -(\dot{M}(x) - 2C(x,\dot{x}))^T. \tag{1.35}$$

The Euler-Lagrange system (1.34) is passive from the generalized torque input τ to the generalized velocity $\dot{x}$. Such a result is established by using (1.35) and taking the total energy of the system $V = K(x,\dot{x}) + P(x) - \bar{P}$ as the storage function. The derivative of V is given by

$$\dot{V} = \dot{x}^T M(x)\ddot{x} + \frac{1}{2}\dot{x}^T \dot{M}(x)\dot{x} + g(x) \tag{1.36}$$

$$= \dot{x}^T \tau + \frac{1}{2}\dot{x}^T (\dot{M}(x) - 2C(x,\dot{x}))\dot{x} \tag{1.37}$$

$$= \dot{x}^T \tau. \tag{1.38}$$

If τ is chosen as

$$\tau = -R\dot{x} + \tau_e, \quad R = R^T > 0, \tag{1.39}$$

we immediately verify the strict output passivity from τ_e to $\dot{x}$. □

Passivity of a linear time invariant dynamic system is closely related to positive realness of the transfer function of that system.

Definition 1.6. [Positive Realness]
A scalar transfer function $g(s)$ is called *positive real* if

- poles of $g(s)$ have nonpositive real parts;
- for all $\omega \in \mathbb{R}$ for which $j\omega$ is not a pole of $g(s)$, $Re[g(j\omega)] \geq 0$;
- any pure imaginary pole $j\omega$ of $g(s)$ is a simple pole and the associated residues are nonnegative. □

The second condition in Definition 1.6 means that the Nyquist plot of $g(j\omega)$ lies in the closed right-half complex plane, which implies that the phase shift of $g(s)$ cannot exceed $\pm 90°$.

Definition 1.7. [Strict Positive Realness [61, 142]]
A transfer function $g(s)$ is called *strictly positive real* if $g(s-\varepsilon)$ is positive real for some $\varepsilon > 0$. □

The strict positive realness of $g(s)$ can also be characterized in the following lemma:

Lemma 1.1. *A scalar transfer function $g(s)$ is strictly positive real if and only if*

- *poles of $g(s)$ have negative real parts;*
- *for all $\omega \in \mathbb{R}$, $Re[g(j\omega)] > 0$;*
- *either $g(\infty) > 0$ or $g(\infty) = 0$ and $\lim_{\omega \to \infty} \omega^2 Re[g(j\omega)] > 0$.* □

Example 1.4. The first-order integrator $g(s) = \frac{1}{s}$ is positive real since it has a simple pole at $\omega = 0$, associated with residue 1, and

$$Re\left[\frac{1}{j\omega}\right] = 0, \quad \forall \omega \neq 0. \tag{1.40}$$

The second-order integrator $g(s) = \frac{1}{s^2}$ is not positive real since the phase shift of $g(s)$ is $-180°$.

The transfer function $g(s) = \frac{1}{as+c}$ for $a, c > 0$ is strictly positive real since $g(s-\varepsilon)$ is positive real for $\varepsilon = c/a > 0$. □

When a transfer function $g(s)$ is realized by a minimal state space representation

$$\mathcal{H}: \quad \begin{cases} \dot{\xi} = A\xi + Bu \\ y = C\xi + Du, \end{cases} \tag{1.41}$$

the positive realness of $g(s)$ means that (1.41) is passive.

Lemma 1.2. *Let $\mathcal{H}$ in (1.41) be a minimal state space representation of $g(s)$. Then,*

- *$\mathcal{H}$ is passive if $g(s)$ is positive real;*
- *$\mathcal{H}$ is strictly passive if $g(s)$ is strictly positive real.* □

The passivity property of a dynamical system remains unchanged when the input and output variables are transformed in a "symmetric" fashion as in Fig. 1.7.

Structure 1 (Symmetric Input-Output Transformation) *Let the system H in Fig. 1.7 be passive and let E be a matrix with a compatible dimension. Then the system in Fig. 1.7 is passive from $\bar{u}$ to $\bar{y}$.* □

Proof. Note that $u^T y = (E^T \bar{u})^T y = \bar{u}^T \bar{y}$. Thus, the passivity from u to y translates to the passivity from $\bar{u}$ to $\bar{y}$. □

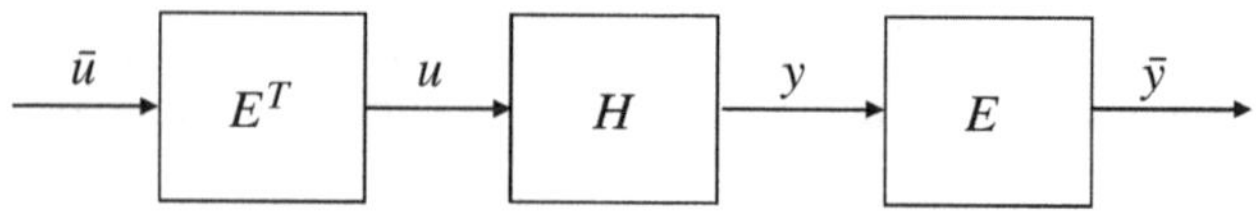

Fig. 1.7 Pre- and post- multiplication of a matrix and its transpose preserves the passivity of H.

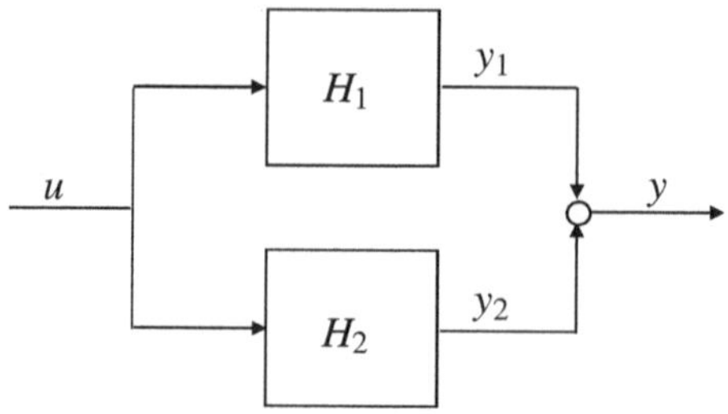

Fig. 1.8 Parallel interconnection of two passive systems.

The definition of passivity closely relates to the stability of (1.26) when $u = 0$. In fact, when the storage function S is positive definite, (1.27) implies that for $u = 0$,

$$\dot{S} \leq -W(\xi) \leq 0. \tag{1.42}$$

Assume that $f(0,0) = 0$. Using standard Lyapunov arguments, we conclude that the unforced system $\dot{\xi} = f(\xi,0)$ has a stable equilibrium $\xi = 0$. If, in addition, $W(\xi)$ is positive definite, $\xi = 0$ is asymptotically stable. If S is also proper, i.e., $S(\xi) \to \infty$ as $|\xi| \to \infty$, the asymptotic stability of $\xi = 0$ is global.

The stability properties are preserved if two or more passive systems are interconnected properly. Among all possible passivity-preserving structures, the following three structures are employed in our cooperative control design.

Structure 2 (Parallel Interconnection) *Consider the parallel interconnection of two passive systems H_1 and H_2 in Fig. 1.8. Then the interconnected system is passive from u to y.* □

Structure 3 (Negative Feedback Interconnection) *Consider the negative feedback interconnection of two passive systems H_1 and H_2 in Fig. 1.9. Then the interconnected system is passive from u to y.* □

Replacing H_1 in Structure 3 with Structure 1, we obtain Structure 4 below:

Structure 4 (Symmetric Interconnection) *Consider the interconnection structure of two passive systems H_1 and H_2 in Fig. 1.10. Then the interconnected system is passive from u to y.* □

We will demonstrate in the next chapter that Structure 4 arises naturally in cooperative control with bidirectional information flow. In particular, the matrices E

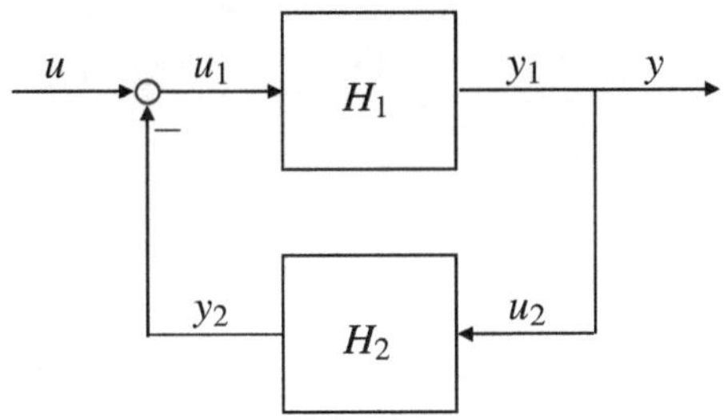

Fig. 1.9 Negative feedback interconnection of two passive systems.

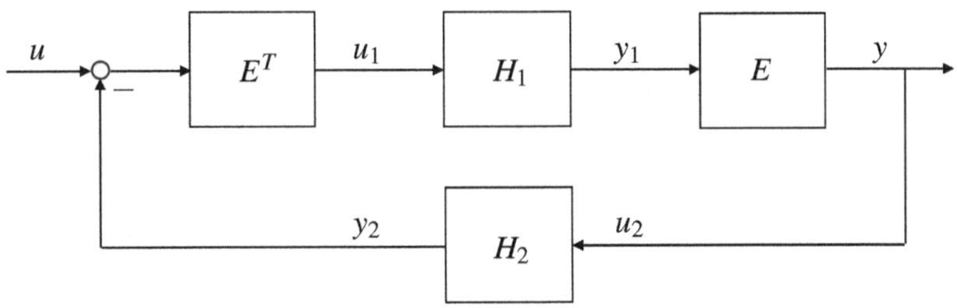

Fig. 1.10 Symmetric Interconnection of two passive systems H_1 and H_2 is still passive.

and E^T are dictated by the undirected information topology between the agents. The H_1 and H_2 blocks in Structure 4, being block diagonal, represent the dynamics of individual agents and their relative configuration, respectively. We will then apply passivation designs to H_1 and H_2 such that the closed-loop stability is guaranteed by Structure 4.

Chapter 2
Passivity As a Design Tool for Cooperative Control

2.1 Introduction

In this chapter, we formulate a coordination problem that is applicable to formation stabilization and group agreement as special cases, and present a class of feedback laws that solve this problem with local information. A key observation is that bidirectional communication gives rise to Structure 4 in Section 1.5, which guarantees that the resulting feedback loop will inherit the *passivity* properties of its components. By exploiting this structure, we develop a design method which results in a broad class of feedback laws that achieve passivity and, thus, stability of the interconnected system. The passivity approach also leads to a systematic construction of a Lyapunov function in the form of a sum of storage functions for the subsystems. As detailed in this chapter, several existing feedback rules for formation stability and group agreement become special cases in the passivity framework.

The coordination task studied in this chapter is to steer the differences between the output variables of group members to a prescribed compact set. This compact set may be a sphere when the outputs are positions of vehicles that must maintain a given distance in a formation, or the origin if the outputs are variables that must reach an agreement across the group. We thus formulate this task as a *set stability* problem and use passivity as a tool for constructing a stabilizing feedback law and a Lyapunov function with respect to this set. We prove global asymptotic stability with additional assumptions that guarantee appropriate *detectability* properties for trajectories away from the set.

2.2 Problem Statement

Consider a group of N agents, where each agent $i = 1, \cdots, N$, is represented by a vector $x_i \in \mathbb{R}^p$ that consists of variables to be coordinated with the rest of the group. The topology of information exchange between these agents is modeled as a graph

G. Since the information flow between neighbors is assumed to be bidirectional, G is an undirected graph. We also assume that G is connected and that G has ℓ undirected links. To simplify the analysis, we assign an orientation to G by considering one of the nodes to be the positive end of the link. As discussed in Section 1.4, the choice of orientation does not change the results because of the symmetric information flow.

The objective is to develop coordination laws that are implementable with local information (agent i can use the information of agent j if agent j is a neighbor) and that guarantee the following *two* group behaviors:
A1) Each agent achieves in the limit a common velocity vector $v(t) \in \mathbb{R}^p$ *prescribed for the group; that is*

$$\lim_{t\to\infty} |\dot{x}_i - v(t)| = 0, \quad i = 1,\cdots,N; \tag{2.1}$$

A2) If agents i and j are connected by link k, then the difference variable z_k

$$z_k := \sum_{l=1}^{N} d_{lk} x_l = \begin{cases} x_i - x_j \text{ if } k \in \mathcal{L}_i^+ \\ x_j - x_i \text{ if } k \in \mathcal{L}_i^- \end{cases} \tag{2.2}$$

converges to a prescribed compact set $\mathcal{A}_k \subset \mathbb{R}^p$, $k = 1,\cdots,\ell$, *where* d_{ik} *is defined in (1.21).*

The reference velocity $v(t)$ can be considered as a "mission plan" of the group. By specifying different $v(t)$, we achieve different group motions, such as rotational and translational motions. Examples of target sets $\mathcal{A}_k$ include the origin if x_i's are variables that must reach an agreement within the group, or a sphere in $\mathbb{R}^p$ if x_i's are positions of vehicles that must maintain a prescribed distance. Objectives A1-A2 may be employed to design and stabilize a formation of vehicles, or to synchronize variables in a distributed network of satellites, etc.

We introduce the concatenated vectors

$$x := [x_1^T,\cdots,x_N^T]^T \in \mathbb{R}^{pN} \quad z := [z_1^T,\cdots,z_\ell^T]^T \in \mathbb{R}^{p\ell}. \tag{2.3}$$

We partition D in terms of columns vectors, i.e.,

$$D = [\, D_1 \,\vdots\, \cdots \,\vdots\, D_\ell \,] \tag{2.4}$$

and note from (2.2) that

$$z_k = (D_k^T \otimes I_p)x. \tag{2.5}$$

Concatenating z_k's together, we have

$$z = (D^T \otimes I_p)x \tag{2.6}$$

which means that z is restricted to be in the range space $\mathcal{R}(D^T \otimes I_p)$. Thus, for the objective A2 to be feasible, the target sets $\mathcal{A}_k$ must satisfy

$$\{\mathcal{A}_1 \times \cdots \times \mathcal{A}_\ell\} \cap \mathcal{R}(D^T \otimes I_p) \neq \emptyset. \tag{2.7}$$

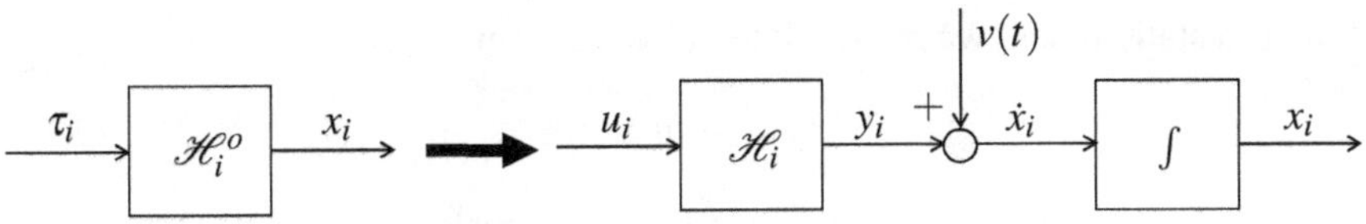

Fig. 2.1 Step 1 transforms agent dynamics from (2.9) to (2.10) by designing an internal feedback τ_i. This internal feedback achieves for agent i passivity from an external feedback signal u_i to the velocity error y_i. The resulting passive block is denoted by $\mathscr{H}_i$.

2.3 The Passivity-based Design Procedure

Step 1. Design an internal feedback *loop for each agent $i = 1,\cdots,N$ that renders its dynamics passive from an external feedback signal u_i (to be designed in Step 2) to the velocity error*

$$y_i := \dot{x}_i - v(t). \tag{2.8}$$

Assume that the input-output dynamics of agent i are given by

$$x_i = \mathscr{H}_i^o\{\tau_i\}, \tag{2.9}$$

where $\mathscr{H}_i^o\{\tau_i\}$ denotes the output of a dynamic system $\mathscr{H}_i^o$ with the control input τ_i. The system $\mathscr{H}_i^o$ may be linear (e.g., single/double integrators) or nonlinear (e.g., Euler-Lagrange equation). In Step 1, we seek a feedback controller τ_i for each agent such that the agent dynamics $\mathscr{H}_i^o$ in (2.9) may be expressed as

$$\dot{x}_i = \mathscr{H}_i\{u_i\} + v(t), \tag{2.10}$$

where $\mathscr{H}_i$ is as in Fig. 2.1. For example, for the first order agent dynamics $\dot{x}_i = \tau_i$, Step 1 is trivially accomplished by choosing $\tau_i = \mathscr{H}_i\{u_i\} + v(t)$.

If $\mathscr{H}_i$ is dynamic, we assume that it is of the form

$$\mathscr{H}_i : \begin{cases} \dot{\xi}_i = f_i(\xi_i, u_i) \\ y_i = h_i(\xi_i, u_i) \end{cases} \tag{2.11}$$

where y_i is the velocity error and $\xi_i \in \mathbb{R}^{n_i}$ is the state variable of subsystem $\mathscr{H}_i$. We assume that $f_i(\cdot,\cdot)$ and $h_i(\cdot,\cdot)$ are C^2 functions such that

$$f_i(0, u_i) = 0 \Rightarrow u_i = 0 \tag{2.12}$$

and

$$h_i(0,0) = 0. \tag{2.13}$$

The main restriction on (2.11) is that it be strictly passive with C^1, positive definite, radially unbounded storage functions $S_i(\xi_i)$ satisfying (1.27) for some positive definite functions $W_i(\cdot)$.

If $\mathscr{H}_i$ is a static block, we restrict it to be of the form

$$y_i = h_i(u_i) \tag{2.14}$$

where $h_i : \mathbb{R}^p \to \mathbb{R}^p$ is a locally Lipschitz function satisfying (1.25) for any $u \neq 0$. In the situation where one of the agents, say agent 1, is the "leader" of the group in the sense that $\dot{x}_1$ uses no feedback term from the other agents, we let

$$h_1(u_1) \equiv 0 \quad \forall u_1 \in \mathbb{R}^p. \tag{2.15}$$

We next show how to apply Step 1 to agents modeled as double integrators. In Chapters 5 and 6, we will demonstrate that broader classes of physical systems, including rigid body rotation and Euler-Lagrange systems in (1.34), may be transformed to the form in Step 1.

Example 2.1 (Step 1 for agents modeled as double integrators).
We consider double integrator agent dynamics

$$m_i \ddot{x}_i = \tau_i, \quad i = 1, \cdots, N \tag{2.16}$$

where m_i is the mass of agent i, $x_i \in \mathbb{R}^p$ denotes the position of agent i and $\tau_i \in \mathbb{R}^p$ is the force input of agent i. For planar agents, $p = 2$ and for spatial agents, $p = 3$.
According to Step 1, we design an internal feedback

$$\tau_i = -k_i(\dot{x}_i - v(t)) + m_i \dot{v}(t) + u_i, \quad k_i > 0 \tag{2.17}$$

which makes use of information available only to agent i itself. This feedback law, together with the change of variables

$$\xi_i = \dot{x}_i - v(t), \tag{2.18}$$

brings (2.16) to the form

$$\dot{x}_i = y_i + v(t) \tag{2.19}$$

$$\mathscr{H}_i : \begin{cases} m_i \dot{\xi}_i &= -k_i \xi_i + u_i \\ y_i &= \xi_i. \end{cases} \tag{2.20}$$

Note that $\mathscr{H}_i$ is first order because we effectively consider ξ_i as the state variable instead of x_i. The transfer matrix of (2.20) from u_i to y_i is

$$H_i(s) = \frac{1}{m_i s + k_i} I_p, \quad k_i > 0, \tag{2.21}$$

which is strictly positive real as shown in Example 1.4. Thus, $\mathscr{H}_i$ is strictly passive due to Lemma 1.2. Indeed, a valid storage function for (2.20) is given by

$$S_i(\xi_i) = \frac{1}{2} m_i \xi_i^T \xi_i. \tag{2.22}$$

It is easy to examine that the assumptions in (2.12) and (2.13) are satisfied in (2.20). Thus, Step 1 is completed with the control law in (2.17). Note that other higher order control laws can be designed to render $\mathscr{H}_i$ strictly passive. □

Step 2. Design an external feedback *signal u_i of the form*

$$u_i = -\sum_{k=1}^{\ell} d_{ik}\psi_k(z_k) \tag{2.23}$$

where z_k's are the relative variables as in (2.2), and the multivariable nonlinearities $\psi_k : \mathbb{R}^p \to \mathbb{R}^p$ are to be designed such that the target sets $\mathscr{A}_k$ are invariant and asymptotically stable.

The external feedback law (2.23) is decentralized and implementable with available information since $d_{ik} \neq 0$ only when link k is connected to node i.

Before specifying the properties of ψ_k, we note from Fig. 2.1 and (2.23) that the interconnection of $\mathscr{H}_i$'s and ψ_k's is as in Fig. 2.2, where

$$u = [u_1^T, \cdots, u_N^T]^T \in \mathbb{R}^{pN} \quad \psi = [\psi_1^T, \cdots, \psi_\ell^T]^T \in \mathbb{R}^{p\ell} \quad y = [y_1^T, \cdots, y_N^T]^T \in \mathbb{R}^{pN}. \tag{2.24}$$

Note from (2.23) that

$$u_i = -[\; d_{i1}I_p \,\vdots\, \cdots \,\vdots\, d_{i\ell}I_p \;]\psi, \tag{2.25}$$

which means

$$u = -(D \otimes I_p)\psi(z). \tag{2.26}$$

Fig. 2.2 exhibits a "symmetric" interconnection structure similar to Structure 4 in Section 1.5. The symmetric interconnection follows from the symmetry inherent in the undirected graphs. This structure allows us to proceed with a passivity-based design of ψ_k, $k = 1, \cdots, \ell$.

We design the nonlinearities $\psi_k(z_k)$ to be of the form

$$\psi_k(z_k) = \nabla P_k(z_k) \tag{2.27}$$

where $P_k(z_k)$ is a nonnegative C^2 function

$$P_k : \mathscr{G}_k \to \mathbb{R}_{\geq 0} \tag{2.28}$$

defined on an open set $\mathscr{G}_k \subseteq \mathbb{R}^p$, where z_k is allowed to evolve. As an illustration, if x_i's are positions of point masses that must maintain a prescribed distance, then the choice $\mathscr{G}_k = \{z_k \mid z_k \in \mathbb{R}^p \setminus 0\}$ disallows the possibility of collisions between linked agents.

To steer z_k's into the target sets $\mathscr{A}_k \subset \mathscr{G}_k$, we let $P_k(z_k)$ and its gradient $\nabla P_k(z_k)$ vanish on $\mathscr{A}_k$, and let $P_k(z_k)$ grow unbounded as z_k goes to the boundary of $\mathscr{G}_k$:

$$P_k(z_k) \to \infty \text{ as } z_k \to \partial\mathscr{G}_k \tag{2.29}$$

$$P_k(z_k) = 0 \Leftrightarrow z_k \in \mathscr{A}_k \tag{2.30}$$

$$\nabla P_k(z_k) = 0 \Leftrightarrow z_k \in \mathscr{A}_k. \tag{2.31}$$

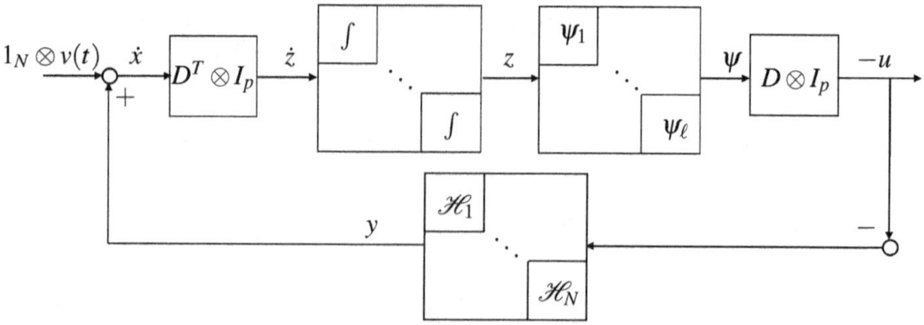

Fig. 2.2 The closed-loop structure of (2.8), (2.11) and (2.26): $\mathscr{H}_i$'s are designed to be strictly passive while pre- and post-multiplying D^T and D preserves the passivity from $\dot{z}$ to ψ. The closed-loop stability follows from the interconnection of two passive systems.

When $\mathscr{G}_k = \mathbb{R}^p$, (2.29) means that $P_k(z_k)$ is radially unbounded. As shown in [137, Remark 2], a continuous function $P_k(z_k)$ satisfying (2.29) and (2.30) exists for any given open set $\mathscr{G}_k$ and compact subset $\mathscr{A}_k \subset \mathscr{G}_k$. We further assume that the sets $\mathscr{A}_k$ and $\mathscr{G}_k$ are chosen such that C^2 smoothness of $P_k(\cdot)$ and (2.31) are also achievable.

For example, if two agents need to reach a common value, we let $\mathscr{A}_k = \{0\}$ and $\mathscr{G}_k = \mathbb{R}^p$. Then the choice of $P_k(z_k) = \frac{1}{2}|z_k|^2$ satisfies (2.28)-(2.31). If two agents must maintain a relative distance of 1, we may choose $\mathscr{A}_k = \{z_k \mid |z_k| = 1\}$ and $\mathscr{G}_k = \{z_k \mid z_k \in \mathbb{R}^p \setminus 0\}$. In this case, a valid choice of P_k is given by $P_k(z_k) = |z_k| - \ln|z_k| - 1$.

The construction of ψ_k as in (2.27) is designed to render the system from $\dot{z}_k$ to ψ_k (and hence from $\dot{z}$ to ψ due to the block diagonal structure in Fig. 2.2) passive. Indeed, consider P_k as a storage function and note that

$$\dot{P}_k = \psi_k(z_k)^T \dot{z}_k, \tag{2.32}$$

which shows the passivity property.

2.4 Stability Results

From Fig. 2.2, the set of equilibria is given by

$$\mathscr{E} = \left\{(z,\xi) \mid \xi = 0,\ (D\otimes I_p)\psi(z) = 0 \text{ and } z \in \mathscr{R}(D^T \otimes I_p)\right\} \tag{2.33}$$

which means that the following property must hold true to ensure that no equilibria arises outside the sets $\mathscr{A}_k$:

Property 2.1. For any $(z,0) \in \mathscr{E}$, i.e., $\xi = 0$, $(D\otimes I_p)\psi(z) = 0$ and $z \in \mathscr{R}(D^T \otimes I_p)$, z satisfies $z \in \mathscr{A}_1 \times \cdots \times \mathscr{A}_\ell$. □

In view of (2.26), Property 2.1 means that

$$u = 0 \Longleftrightarrow z = (D^T \otimes I_p)x \in \mathscr{A}_1 \times \ldots \times \mathscr{A}_\ell. \tag{2.34}$$

When the graph has no cycles, that is, when the columns of D are linearly independent, then $(D \otimes I_p)\psi(z) = 0$ implies $\psi(z) = 0$ and $z_k \in \mathscr{A}_k$ follows from (2.31). Thus, Property 2.1 holds for acyclic graphs. When the columns of D are linearly dependent, whether Property 2.1 holds depends on the sets $\mathscr{A}_k$ and ψ_k. As we will illustrate, it holds in agreement problems where $\mathscr{A}_k$ is the origin but fails in the formation control problem with distance only criterion where $\mathscr{A}_k$ is a sphere.

The feedback interconnection shown in Fig. 2.2 is of the same form as Structure 4 in Section 1.5. The storage functions for the feedforward and feedback subsystems of Fig. 2.2 are, respectively,

$$V_f(z) := \sum_{i=1}^{\ell} P_k(z_k) \quad \text{and} \quad V_b(\xi) := \sum_{i \in \mathscr{I}} S_i(\xi_i) \tag{2.35}$$

where $\mathscr{I}$ denotes the subset of indices $i = 1, \cdots, N$ that correspond to dynamic blocks $\mathscr{H}_i$. In particular, the passivity of the feedforward subsystems follows from Structure 1 in Section 1.5. Using the passivity of the feedforward and feedback subsystems and Structure 4 in Section 1.5 and imposing Property 2.1, we prove asymptotic stability of the set of points where $\xi = 0$ and $z_k \in \mathscr{A}_k$ by taking as a Lyapunov function the sum of the two storage functions in (2.35). This construction results in a Lur'e-type Lyapunov function because its key ingredient $P_k(z_k)$ is the integral of the feedback nonlinearity $\psi_k(z_k) = \nabla P_k(z_k)$. We summarize the main stability result in the following theorem.

Theorem 2.1. *Consider the closed-loop system (2.8), (2.11) and (2.23), where $v(t)$ is uniformly bounded and piecewise continuous and $\mathscr{H}_i$, $i = 1, \cdots, N$, and ψ_k, $k = 1, \cdots, \ell$ are designed as in (2.11)-(2.15) and (2.27)-(2.31) for given open sets $\mathscr{G}_k \subseteq \mathbb{R}^p$ and compact subsets $\mathscr{A}_k \subset \mathscr{G}_k$, where $\mathscr{A}_k$ are as in (2.7). Then:*
i) The feedforward path in Fig. 2.2 is passive from $\dot{x}$ to $-u$, and from y to u;
ii) The feedback path is passive from input u to y;
iii) When Property 2.1 holds, the set

$$\mathscr{A} = \{(z, \xi) \mid \xi = 0, z \in \mathscr{A}_1 \times \cdots \times \mathscr{A}_\ell \cap \mathscr{R}(D^T \otimes I_p)\} \tag{2.36}$$

is uniformly asymptotically stable with region of attraction

$$\mathscr{G} = \{(z, \xi) \mid \xi \in \mathbb{R}^{n_1} \times \cdots \times \mathbb{R}^{n_N}, z \in \mathscr{G}_1 \times \cdots \times \mathscr{G}_\ell \cap \mathscr{R}(D^T \otimes I_p)\}. \tag{2.37}$$

Moreover, all trajectories $(z(t), \xi(t))$ starting in $\mathscr{G}$ converge to the set of equilibria $\mathscr{E}$ in (2.33). □

When Property 2.1 fails, Theorem 2.1 proves that all trajectories converge to the set of equilibria $\mathscr{E}$ in (2.33). In this case, it is possible to conclude "generic convergence" to $\mathscr{A}$ from almost all initial conditions if one can show that the equilibria outside of $\mathscr{A}$ are unstable. We will illustrate such an example in Section 2.7.

Convergence to $\mathscr{A}$ means that the difference variables z_k tend to the target sets $\mathscr{A}_k$. It also implies that $\xi = 0$, $u = 0$ and thus, y in (2.8) is zero, which means that both objectives A1 and A2 are indeed achieved.

Proof (Proof of Theorem 2.1).

i) To prove passivity from $\dot{x}$ to $-u$ we use $V_f(z)$ in (2.35) as a storage function, and obtain from (2.27), (2.6), (2.26) and $(D \otimes I_p)^T = D^T \otimes I_p$:

$$\dot{V}_f = \psi^T \dot{z} = \psi^T (D^T \otimes I_p)\dot{x} = \{(D \otimes I_p)\psi\}^T \dot{x} = -u^T \dot{x}. \tag{2.38}$$

To show passivity from y to $-u$ we substitute $\dot{x} = 1_N \otimes v(t) + y$ in (2.38) and use the fact $(D^T \otimes I_p)(1_N \otimes v(t)) = 0$ from the third item in Property 1.5, thus obtaining

$$\begin{aligned}\dot{V}_f &= \psi^T (D^T \otimes I_p)\{1_N \otimes v(t) + y\} = \psi^T (D^T \otimes I_p) y \\ &= \{(D \otimes I_p)\psi\}^T y = -u^T y.\end{aligned} \tag{2.39}$$

ii) To establish passivity of the feedback path, we let $\mathscr{I}$ denote the subset of indices $i = 1, \cdots, N$ for which $\mathscr{H}_i$ is a dynamic block as in (2.11), and employ the storage function $V_b(\xi)$ in (2.35), which yields:

$$\dot{V}_b = \sum_{i \in \mathscr{I}} \dot{S}_i \leq \sum_{i \in \mathscr{I}} (-W_i(\xi_i) + u_i^T y_i). \tag{2.40}$$

Adding to the right-hand side of (2.40)

$$\sum_{i \notin \mathscr{I}} u_i^T y_i \geq 0 \tag{2.41}$$

which is nonnegative because the static blocks satisfy (1.25) or (2.15), we get

$$\dot{V}_b \leq \sum_{i \in \mathscr{I}} (-W_i(\xi_i) + u_i^T y_i) + \sum_{i \notin \mathscr{I}} u_i^T y_i \leq -\sum_{i \in \mathscr{I}} W_i(\xi_i) + u^T y \tag{2.42}$$

and, thus, conclude passivity with input u and output y.

iii) To prove asymptotic stability of the set $\mathscr{A}$ we use the Lyapunov function

$$V(z, \xi) = V_f(z) + V_b(\xi) \tag{2.43}$$

which is zero on the set $\mathscr{A}$ due to property (2.30), and grows unbounded as (z, ξ) approaches $\partial \mathscr{G}^\infty$ due to property (2.29). From (2.39), (2.40) and (2.41), this Lyapunov function yields the negative semidefinite derivative

$$\dot{V} \leq -\sum_{i \in \mathscr{I}} W_i(\xi_i) - \sum_{i \notin \mathscr{I}} u_i^T y_i, \tag{2.44}$$

which implies that the trajectories $(z(t), \xi(t))$ are bounded on $t \in [0, T]$, for any T within the maximal interval of definition $[0, t_f)$ for the differential equations (2.8), (2.11). Because this bound does not depend on T, and because $v(t)$ is bounded, from

(2.8) we can find a bound on $x(t)$ that grows linearly in T. This proves that there is no finite escape time because, if t_f were finite, by letting $T \to t_f$ we would conclude that $x(t_f)$ exists, which is a contradiction.

Having proven the existence of solutions for all $t \geq 0$, we conclude from (2.44) stability of the set $\mathscr{A}$. However, because the right-hand side of (2.44) vanishes on a superset of $\mathscr{A}$, to prove attractivity of $\mathscr{A}$ we appeal to the Invariance Principle[1] reviewed in Appendix B.2. To investigate the largest invariant set where $\dot{V}(z,\xi) = 0$ we note from (2.12) that if $\xi_i = 0$ holds identically then $u_i = 0$. Likewise the static blocks satisfy (1.25) or (2.15), which means that the right-hand side of (2.44) vanishes when $u_i = 0$, $i = 1, \cdots, N$. Indeed, if the first member $i = 1$ satisfies (1.25), then $u_1 = 0$ follows directly. If it satisfies (2.15) instead of (1.25), $u_1 = 0$ still holds because the sum of the rows of D being zero implies, from (2.6), that

$$u_1 = -\sum_{i=2}^{N} u_i = 0. \tag{2.45}$$

We thus conclude that $u = 0$, which means from (2.26) that $\psi(z)$ lies in the null space $\mathscr{N}(D \otimes I_p)$. Using the Invariance Principle, which states that all bounded solutions approach their positive limit set, which is invariant, we conclude that the trajectories converge to the set $\mathscr{E}$ in (2.33). When Property 2.1 holds, $\mathscr{E}$ coincides with $\mathscr{A}$, which proves asymptotic stability of $\mathscr{A}$ with region of attraction $\mathscr{G}$, while uniformity of asymptotic stability follows from the time-invariance of the (z,ξ)-dynamics. □

The Lyapunov function $V(z,\xi)$ in the proof above yields a negative semidefinite derivative. By using the observability condition in (2.12), we prove the stability results in Theorem 2.1. This Lyapunov function allows us to develop different adaptive schemes to enhance robustness of group motion. For example, in Chapters 3 and 4, we develop adaptive schemes that enable agents to estimate leader's mission plan $v(t)$. These adaptive schemes relax the assumption in Theorem 2.1 that all the agents must have the $v(t)$ information. In Chapter 6, where agreement of multiple Euler-Lagrange systems is studied, we attempt to design adaptive control laws from $V(z,\xi)$ to compensate for uncertainties in Euler-Lagrange systems. However, we illustrate with an example that the resulting adaptive design is not sufficient to ensure group objectives. This is because $\dot{V}$ is only negative *semidefinite*. We will detail in Chapter 6 how we overcome this insufficiency by exploiting the structure of Euler-Lagrange equations and the design flexibility offered by the passivity-based framework.

[1] The Invariance Principle is indeed applicable because the dynamics of (z,ξ) are autonomous: Although $v(t)$ appears in the block diagram in Fig. 2.2, it is canceled in the $\dot{z}$ equation because $(D^T \otimes I_p)(1_N \otimes v(t)) = 0$.

2.5 Application to the Agreement Problem

In several cooperative tasks, it is of interest to steer group variables, such as position, heading, phase of oscillators, to a common value. To apply Theorem 2.1 to this problem, we let $x_i \in \mathbb{R}^p$ denote a vector of variables of interest, and select the target sets to be $\mathscr{A}_k = \{0\}$. With this choice of $\mathscr{A}_k$, the target set constraint (2.7) is trivially satisfied. We may choose $P_k(z_k)$ as a positive definite, radially unbounded function on $\mathscr{G}_k = \mathbb{R}^p$ with the property

$$z_k^T \nabla P_k(z_k) = z_k^T \psi_k(z_k) > 0 \quad \forall z_k \neq 0 \tag{2.46}$$

so that (2.27)-(2.31) and Property 2.1 hold. In particular, Property 2.1 holds because $z \in \mathscr{R}(D^T \otimes I_p)$ and $\psi(z) \in \mathscr{N}(D \otimes I_p)$ imply that z and $\psi(z)$ are orthogonal to each other, which, in view of (2.46), is possible only if $z = 0$.

Corollary 2.1. *Consider agents $i = 1, \cdots, N$, interconnected as described by the graph representation (1.21), and let z_k, $k = 1, \cdots, \ell$ denote the differences between the variables x_i of neighboring members as in (2.2). Let $P_k(z_k)$ be positive definite, radially unbounded functions satisfying (2.46) and let $\psi_k(z_k) = \nabla P_k(z_k)$. Then the agreement protocol*

$$\dot{x}_i = \mathscr{H}_i \left\{ -\sum_{i=1}^{\ell} d_{ik} \psi_k(z_k) \right\} + v(t), \quad i = 1, \cdots, N \tag{2.47}$$

where $\mathscr{H}_i\{u_i\}$ denotes the output at time t of a static or dynamic block satisfying (2.11)-(2.15), guarantees $|\dot{x}_i - v(t)| \to 0$ and

$$x_i - x_j \to 0 \quad as \quad t \to \infty \tag{2.48}$$

for every pair of nodes (i, j) which are connected by a path. □

When $p = 1$, that is when x_i's and z_k's are scalars, condition (2.46) means that $\psi_k(z_k) = \nabla P_k(z_k)$ is a *sector* nonlinearity which lies in the first and third quadrants. Corollary 2.1 thus encompasses the result of [102], which proposed agreement protocols of the form

$$\dot{x}_i = - \sum_{j \in \mathscr{N}_i} \phi_{ij}(x_i - x_j) \tag{2.49}$$

where $\phi_{ij}(\cdot) = \phi_{ji}(\cdot)$ plays the role of our $\psi_k(\cdot)$. However, both [102] and a related result in [122] assume that the nonlinearities $\phi_{ij}(\cdot)$ satisfy an *incremental sector* assumption which is more restrictive than the sector condition (2.46) of Corollary 2.1. An independent study in [147] takes a similar approach to synchronization as [122]; however, it further restricts the coupling terms $\phi_{ij}(\cdot)$ to be linear. The feedback law (2.47) in Corollary 2.1 generalizes (2.49) by applying to its right-hand side the additional operation $\mathscr{H}_i\{\cdot\}$, which may either represent a passive filter or another sector nonlinearity $h_i(\cdot)$ as specified in Section 2.3. Because $\mathscr{H}_i$ in (2.47) can be dynamic, Corollary 2.1 is applicable, unlike other agreement results surveyed in [110], to plants with higher-order dynamics than an integrator. See, for example, the second order system in Section 2.6.1.

2.6 Position-based Formation Control As a Shifted Agreement Problem

One of the major topics in cooperative control is the formation maintenance and stability, where the goal is to drive relative positions (i.e., z_k's) or relative distances (i.e., $|z_k|$'s) between agents to prescribed values. Depending on the goal, we may pursue one of the following:

- *distance-based formation control*, where the desired target set $\mathcal{A}_k$ in objective A2 is given by
$$\mathcal{A}_k = \{z_k \mid |z_k| = d_k\}, \quad d_k \in \mathbb{R}_{>0},\ k = 1, \cdots, \ell; \tag{2.50}$$
- *position-based formation control*, where the desired target set $\mathcal{A}_k$ in objective A2 is given by
$$\mathcal{A}_k = \{z_k \mid z_k = z_k^d\}, \quad z_k^d \in \mathbb{R}^p,\ k = 1, \cdots, \ell. \tag{2.51}$$

The goal of the distance-based formation control is to achieve a desired shape of the group formation while the position-based formation control concerns not only the desired shape but also the desired orientation of the group formation. We first consider the position-based formation control and demonstrate that it can be transformed to an agreement problem.

The set points z_k^d in (2.51) dictate the relative configuration of the group. When the graph contains cycles, the sum of the relative position vectors z_j over each cycle must be zero; that is, $z = [z_1^T, \cdots, z_\ell^T]^T$ must lie in the range space of $D^T \otimes I_p$ so that (2.7) holds. We thus assume that $z^d = [(z_1^d)^T, \cdots, (z_\ell^d)^T]^T$ is designed to lie in the range space of $D^T \otimes I_p$, which means that

$$z^d = (D^T \otimes I_p)x_c \tag{2.52}$$

for some $x_c \in \mathbb{R}^{pN}$. The condition (2.52) implies that (2.7) is satisfied.

Introducing

$$\mathbf{x}(t) := x(t) - x_c - \int_0^t 1_N \otimes v(\tau) d\tau, \tag{2.53}$$

where x_c is as in (2.52), and

$$\mathbf{z} = (D^T \otimes I_p)\mathbf{x} = z - z^d, \tag{2.54}$$

we notice that objectives A1-A2 for the position-based formation control translate to the asymptotic stability of the origin for

$$X = [\dot{\mathbf{x}}^T \ \ \mathbf{z}^T]^T. \tag{2.55}$$

According to Corollary 2.1, the global asymptotic stability of $X = 0$ is guaranteed by the protocol

$$\dot{\mathbf{x}}_i = \mathscr{H}_i\left\{-\sum_{i=1}^{\ell} d_{ik}\psi_k(\mathbf{z}_k)\right\}, \quad i = 1, \cdots, N \tag{2.56}$$

where $\psi_k(\cdot)$ satisfies (2.46). Using (2.23), (2.53) and (2.54), we rewrite (2.56) in the original coordinate $(\dot{x}, z)$ as

$$\dot{x}_i = y_i + v(t) \tag{2.57}$$
$$y_i = \mathscr{H}_i\{u_i\} \tag{2.58}$$

where

$$u_i = -\sum_{i=1}^{\ell} d_{ik}\psi_k(z_k - z_k^d). \tag{2.59}$$

Corollary 2.2. *Consider a group of agents $i = 1, \cdots, N$. The protocol (2.57)-(2.58), where $\psi_k = \nabla P_k(z_k)$ satisfies (2.46), guarantees that*

$$|\dot{x}_i - v(t)| \to 0, \quad \forall i, \tag{2.60}$$

and

$$z_k \to z_k^d, \quad \forall k. \tag{2.61}$$

□

Example 2.2 (Collision avoidance).

The closed-loop system (2.57)-(2.58) ensures only the convergence to the desired formation. Other objectives, such as collision avoidance, can be achieved by incorporating additional feedback terms. For example, to avoid collision, we employ the artificial potential field approach in robotics and augment u_i in (2.58) as

$$u_i = -\sum_{k=1}^{\ell} d_{ik}\psi_k(z_k - z_k^d) - \sum_{j=1}^{N} \nabla_{x_i} Q_{ij}(|x_i - x_j|) \tag{2.62}$$

where the C^1 artificial potential function $Q_{ij}(\cdot) : \mathbb{R}_{\geq 0} \to \mathbb{R}_{\geq 0}$ satisfies

$$Q_{ij}(s) \to \infty \quad \text{as} \quad s \to 0 \tag{2.63}$$
$$Q_{ij}(s) = 0 \quad \text{as} \quad s > R \tag{2.64}$$

for some positive R. Using the Lyapunov function

$$V = \sum_{i=1}^{N} S_i(\xi_i) + \sum_{k=1}^{\ell} P_k(z_k - z_k^d) + \sum_{i=1}^{N}\sum_{j>i} Q_{ij}(|x_i - x_j|) \tag{2.65}$$

we obtain

$$\dot{V} \leq -\sum_{i=1}^{N} W_i(\xi) - \sum_{i=1}^{N}\left(y_i^T \sum_{j=1, j\neq i}^{N} \nabla_{x_i} Q_{ij}(|x_i - x_j|)\right)$$
$$+ \sum_{i=1}^{N}\sum_{j>i}\left((\nabla_{x_i} Q_{ij}(|x_i - x_j|))^T \dot{x}_i + (\nabla_{x_j} Q_{ij}(|x_i - x_j|))^T \dot{x}_j\right). \quad (2.66)$$

Since $\nabla_{x_i} Q_{ij}(|x_i - x_j|) = -\nabla_{x_j} Q_{ij}(|x_i - x_j|)$, we rewrite $\dot{V}$ using (2.57) as

$$\dot{V} \leq -\sum_{i=1}^{N} W_i(\xi) - \sum_{i=1}^{N}\sum_{j=1, j\neq i}^{N} y_i^T (\nabla_{x_i} Q_{ij}(|x_i - x_j|))$$
$$+ \sum_{i=1}^{N}\sum_{j>i}\left((\nabla_{x_i} Q_{ij}(|x_i - x_j|))^T y_i + (\nabla_{x_j} Q_{ij}(|x_i - x_j|))^T y_j\right)$$
$$= -\sum_{i=1}^{N} W_i(\xi) - \sum_{i=1}^{N}\sum_{j<i}^{N} y_i^T (\nabla_{x_i} Q_{ij}(|x_i - x_j|)) + \sum_{i=1}^{N}\sum_{j>i} (\nabla_{x_j} Q_{ij}(|x_i - x_j|))^T y_j$$
$$= -\sum_{i=1}^{N} W_i(\xi) \leq 0. \quad (2.67)$$

Thus, V in (2.65) is nonincreasing, that is, $V(t) \leq V(0)$. Since $V \to \infty$ as $|x_i - x_j| \to 0$, $\forall i \neq j$, the boundedness of $V(t)$ implies collision avoidance.

Applying the Invariance Principle, we conclude from (2.67) that $\xi \to 0$, which implies from (2.12) that $u \to 0$. Note that due to the additional term in (2.62) that handles collision avoidance, $u_i \to 0$ does not mean $z_k \to z_k^d$, that is, convergence to the desired formation is not guaranteed. Indeed, there may exist an asymptotically stable equilibrium where $u_i = 0$ and the desired formation is not achieved. This equilibrium corresponds to a local minimum of the potential function V in (2.65). To eliminate such a local minima, one may apply navigation function techniques in [113] to the construction of P_k and Q_{ij} such that from almost all initial conditions, the agents converge to the desired formation. We refer interested readers to [133, 132] for further details on applying navigation function to formation control. □

2.6.1 Design Example

We consider a group of agents modeled as (2.16) in Example 2.1. The feedback law (2.17) achieves Step 1. We next apply Step 2 and design u_i. According to Corollary 2.2, we take

$$u_i = -\sum_{i=1}^{\ell} d_{ik} \psi_k(z_k - z_k^d), \quad (2.68)$$

where $\psi_k(\cdot)$ satisfies (2.46). The closed-loop system of (2.19), (2.20) and (2.59) is given by

$$m_i(\ddot{x}_i - \dot{v}(t)) + k_i(\dot{x}_i - v(t)) + \sum_{i=1}^{\ell} d_{ik}\psi_k(z_k - z_k^d) = 0 \tag{2.69}$$

which can be rewritten as

$$(M \otimes I_p)\ddot{\mathbf{x}} + (K \otimes I_p)\dot{\mathbf{x}} + (D \otimes I_p)\psi(\mathbf{z}) = 0 \tag{2.70}$$

where $M = \text{diag}\{m_1, \cdots, m_N\}$ and $K = \text{diag}\{k_1, \cdots, k_N\}$.

We now show that for quadratic potential function P_k, (2.70) recovers a second order linear consensus protocol. To this end, we choose a quadratic potential function

$$P_k = \frac{\delta_k}{2}|z_k - z_k^d|^2 \quad \delta_k \in \mathbb{R}_{>0} \tag{2.71}$$

which leads to

$$\psi_k(z_k) = \delta_k(z_k - z_k^d). \tag{2.72}$$

The constants δ_k's are the feedback gains which regulate the relative emphasis of $|z_k - z_k^d|$ for different k's. Defining

$$\Delta = \text{diag}\{\delta_1, \cdots, \delta_\ell\} \tag{2.73}$$

and substituting (2.54) in (2.70), we obtain

$$(M \otimes I_p)\ddot{\mathbf{x}} + (K \otimes I_p)\dot{\mathbf{x}} + (L_\Delta \otimes I_p)\mathbf{x} = 0 \tag{2.74}$$

where $L_\Delta = D\Delta D^T$ is the *weighted* graph Laplacian (recall that without the subscript "Δ", L denotes the *unweighted* Laplacian $L = DD^T$). The closed-loop system (2.74) is a second order linear consensus protocol well studied in the literature (see e.g., [109]). The design in (2.70) gives a passivity interpretation of the second order consensus protocol (2.74) and extends it to nonlinear coupling ψ_k.

Example 2.3 (Agreement of Second-order Agents with Directed Graphs).

When the graph is undirected, the stability of (2.74) holds for arbitrary $m_i > 0$ and $k_i > 0$, $\forall i$. For directed graphs, however, (2.74) may become unstable even for uniform m_i and k_i. To illustrate this, let us take $p = 1$ (scalar variables), $m_i = 1$, $k_i = 1$, and $\delta_k = 1$, $\forall i$, $\forall k$, in (2.74), which leads to

$$\ddot{\mathbf{x}} + \dot{\mathbf{x}} + L\mathbf{x} = 0 \tag{2.75}$$

where L is defined in (1.11).

To investigate the stability of (2.75), we use the Schur decomposition reviewed in Appendix B.1 and decompose L as

$$L = QBQ^{-1} \tag{2.76}$$

where Q is a unitary complex matrix and B is an upper triangular matrix with all the eigenvalues of L on the diagonal of B. Note that if L is symmetric, i.e., the graph G is undirected, Q can chosen as the orthonormal eigenvectors of L and accordingly B

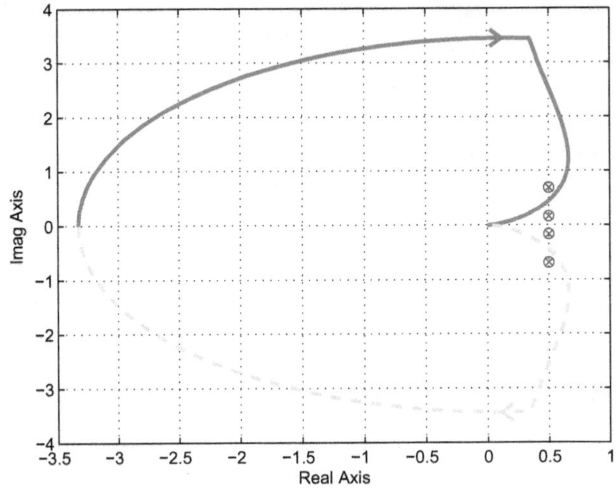

Fig. 2.3 Nyquist plot of $-\frac{1}{s(s+1)}$. Four ⊗'s denote the inverse of four nonzero eigenvalues of the graph Laplacian matrix when the information topology is a directed cycle of 5 agents.

is a diagonal matrix. We will use this decomposition technique again in Chapter 9 to study robustness properties of (2.74) with undirected graphs.

Using a coordinate transformation

$$d = Q^{-1}\mathbf{x}, \tag{2.77}$$

we obtain from (2.75)

$$\ddot{d} + \dot{d} + Bd = 0, \quad i = 1, \cdots, N. \tag{2.78}$$

Because B is upper triangular and because the eigenvalues of L are the diagonal elements of B, the stability of (2.78) is equivalent to the stability of

$$\ddot{d} + \dot{d} + \lambda_i d = 0, \quad i = 1, \cdots, N, \tag{2.79}$$

where λ_i is the ith eigenvalue of L. If $\lambda_i = 0$ for some i, (2.79) is stable. It then follows that (2.79) (and thus (2.75)) is stable if and only if the Nyquist plot of $-\frac{1}{s(s+1)}$ does not encircle λ_i^{-1} for any nonzero λ_i.

The Nyquist plot of $-\frac{1}{s(s+1)}$ is shown by the solid-dash line in Fig. 2.3. For undirected graphs, λ_i is nonnegative and the Nyquist plot never encircles nonnegative real axis. Therefore, the stability of (2.75) is guaranteed independently of the graph and the number of the agents. However, for directed graphs, λ_i may become complex and thus the graph and the number of agents may affect stability. For example, consider a directed cyclic graph of N agents, where agent i is the only neighbor of

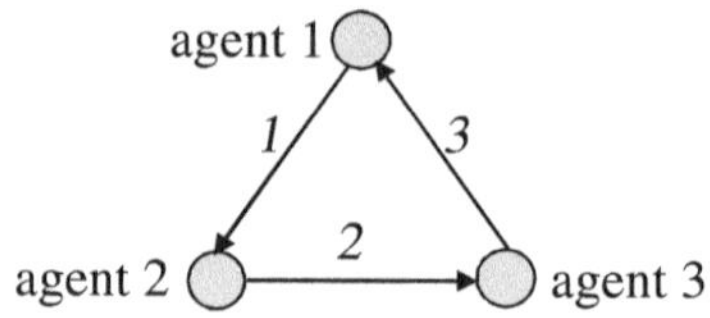

Fig. 2.4 We assign an orientation to an undirected graph of three agents, where every two agents are neighbors. The link number is put beside each link.

agent $i+1$, $i = 1,\cdots,N-1$, and agent N is the only neighbor of agent 1. For $N < 5$, (2.75) is stable. However, for $N = 5$, there exists two λ_i such that λ_i^{-1} is encircled by the Nyquist plot of $-\frac{1}{s(s+1)}$, as shown in Fig. 2.3. Thus, (2.75) becomes unstable, which implies that for directed graphs, closed-loop stability is sensitive to the graph structure and to the number of agents. □

2.6.2 A Simulation Example

In this section, we simulate the position-based formation control system (2.74) and demonstrate that different orientations of the formation can be achieved by modifying $\mathscr{A}_k$'s.

We consider a group of three planar agents (i.e., $p = 2$), where any two agents are neighbors. As shown in Fig. 2.4, we define

$$z_1 = x_2 - x_1, \quad z_2 = x_3 - x_2, \quad z_3 = x_1 - x_3 \tag{2.80}$$

and design desired target sets for z_k's to be

$$\begin{aligned}
\mathscr{A}_1 &= \left\{ z_1 \middle| z_1 = z_1^d = [-\tfrac{\sqrt{3}}{2} \ \tfrac{1}{2}]^T \right\}, \\
\mathscr{A}_2 &= \left\{ z_2 \middle| z_2 = z_2^d = [0 \ -1]^T \right\}, \\
\mathscr{A}_3 &= \left\{ z_3 \middle| z_3 = z_3^d = [\tfrac{\sqrt{3}}{2} \ \tfrac{1}{2}]^T \right\}.
\end{aligned} \tag{2.81}$$

We choose $M = \text{diag}\{5,2,1\}$ and $K = 5I_3$ in (2.74). The reference velocity $v(t)$ is zero. The weight Δ in (2.73) is set to I_3. The initial positions of the agents are $x_1(0) = [5\ 0]^T$, $x_2(0) = [2\ 2]^T$, and $x_3(0) = [0\ 0]^T$. Simulation result in Fig. 2.5 shows that the desired formation is achieved.

We now modify the desired target sets in (2.81) to

$$\begin{aligned}
\mathscr{A}_1 &= \left\{ z_1 \middle| z_1 = z_1^d = [-\tfrac{1}{2} \ -\tfrac{\sqrt{3}}{2}]^T \right\}, \\
\mathscr{A}_2 &= \left\{ z_2 \middle| z_2 = z_2^d = [1 \ 0]^T \right\}, \\
\mathscr{A}_3 &= \left\{ z_3 \middle| z_3 = z_3^d = [-\tfrac{1}{2} \ \tfrac{\sqrt{3}}{2}]^T \right\}.
\end{aligned} \tag{2.82}$$

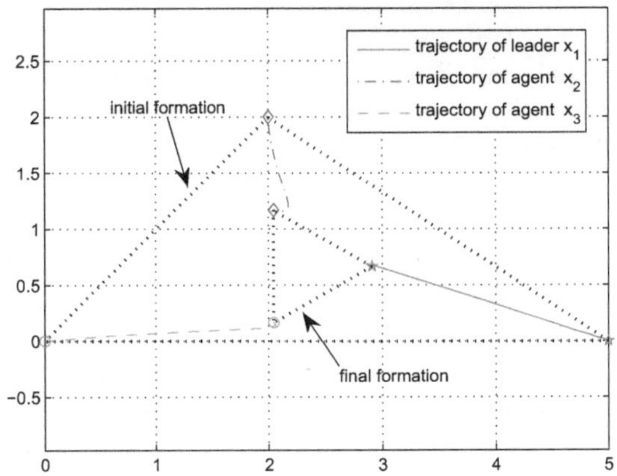

Fig. 2.5 Using (2.70) and (2.72), three agents converge to the desired formation specified in (2.81). Agents 1, 2 and 3 are denoted $\star$, $\diamond$, and $\circ$, respectively.

As shown in Fig. 2.6, (2.82) corresponds to the desired formation in Fig. 2.5 rotated counterclockwise by 90 degrees.

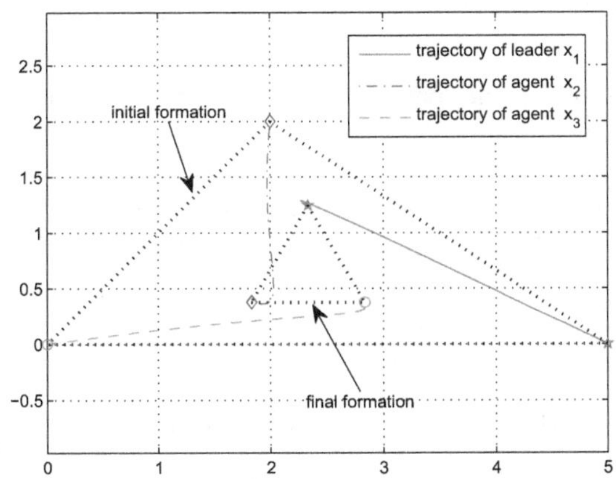

Fig. 2.6 Using (2.70) and (2.72), three agents converge to the desired formation specified in (2.82). Agents 1, 2 and 3 are denoted $\star$, $\diamond$, and $\circ$, respectively.

We see that the position-based formation control stabilizes both the shape and the orientation of the group formation. If the shape of the group formation is the only concern, the distance-based formation control studied in the next section is more appropriate.

2.7 Distance-based Formation Control

In this section, we study the distance-based formation control problem defined in (2.50). In contrast to the position-based formation control, this problem concerns only the shape of the group formation. It will become clear that a key complication is that Property 2.1, which holds true in the position-based formation control, is no longer satisfied in the distance-based formation control with cyclic graphs. Thus, additional undesired equilibria may arise due to the cycles in the graph, making global stabilization of the desired formation impossible. In the special case of three agents, we show that the undesired equilibria are unstable. We then conclude generic convergence to the desired formation from almost all initial conditions.

We also explore existence and uniqueness of the formation shape in this section. The existence of a formation shape is related to the requirement in (2.7), which is further sharpened to sufficient conditions on the desired target sets. These conditions are generalizations of the triangle inequality. We explore the uniqueness issue using a four-agent example. If the shape of the desired formation is a rectangle, specifying desired relative distances of the four sides is not sufficient since the agents may reach a parallelogram instead. In this case, desired relative distances of the diagonal links must also be specified to ensure that the rectangle shape is the unique desired formation.

2.7.1 Passivity-based Design

We assume that Step 1 of the passivity-based design has been achieved. We now proceed to Step 2 and design the nonlinearities ψ_k's. The control objective is to stabilize a formation where the relative distances $|z_k|$, $k = 1, \cdots, \ell$, are equal to $d_k > 0$. We choose $\mathcal{G}_k$ to be $\mathbb{R}^p \setminus \{0\}$ and let the potential functions P_k be a function of z_k satisfying (2.27)-(2.31). An example of $P_k(z_k)$ is given by

$$P_k(z_k) = \int_{d_k}^{|z_k|} \sigma_k(s)ds \tag{2.83}$$

where $\sigma_k : \mathbb{R}_{>0} \to \mathbb{R}$ is a C^1, strictly increasing function such that

$$\sigma_k(d_k) = 0, \tag{2.84}$$

and such that, as $|z_k| \to 0$ and as $|z_k| \to \infty$, $P_k(z_k) \to \infty$ in (2.83). An illustration of $P_k(z_k)$ is shown in Fig. 2.7, where

$$\sigma_k(s) = \frac{1}{d_k} - \frac{1}{s}, \tag{2.85}$$

$$P_k(z_k) = \frac{|z_k|}{d_k} - \ln \frac{|z_k|}{d_k} - 1 \tag{2.86}$$

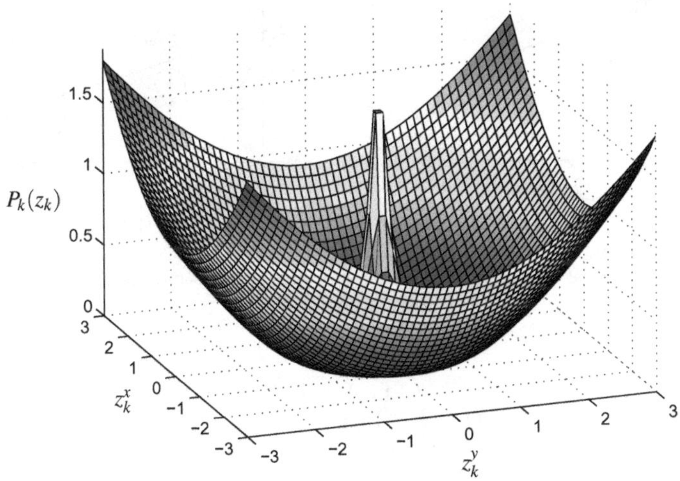

Fig. 2.7 The shape of $P_k(z_k)$: The minima of $P_k(z_k) = |z_k| - \ln|z_k| - 1$ occur on the unit circle $|z_k| = d_k = 1$. The peak at the origin guarantees the collision avoidance between the linked agents.

and d_k is set to 1. Note that the condition $P_k(z_k) \to \infty$ as $|z_k| \to 0$ is imposed only to ensure collision avoidance between linked agents. Since $P_k(z_k)$ satisfies (2.27)-(2.31), the feedback law u_i in (2.23) with the interaction forces

$$\psi_k(z_k) = \nabla P_k(z_k) = \sigma_k(|z_k|)\frac{1}{|z_k|}z_k \quad z_k \neq 0 \tag{2.87}$$

guarantees global asymptotic stability of the desired formation from Theorem 2.1 when the graph G is acyclic.

For cyclic graphs, we need to examine whether or not Property 2.1 is satisfied. We consider an example of three agents, where each agent is a neighbor of the other two agents, thereby forming a cycle in the graph G. Let the desired formation be an equilateral triangle shown in 2.8(a) with $d_k = 1$, $k = 1,2,3$. Note that $\psi_k(z_k)$ in (2.87) plays the role of a "spring force" which creates an attraction force when $|z_k| > 1$ and a repulsion force when $|z_k| < 1$. When $u_i = 0$, additional equilibria arise when the point masses are aligned as in Fig. 2.8(b), and the attraction force between the two distant masses counterbalances the repulsion force due to the intermediate mass.

To characterize such equilibria, we let the middle agent in Fig. 2.8(b) be agent 2 and define

$$z_1 = x_1 - x_2 \quad z_2 = x_2 - x_3 \quad \text{and} \quad z_3 = x_1 - x_3, \tag{2.88}$$

which implies from (2.6) that

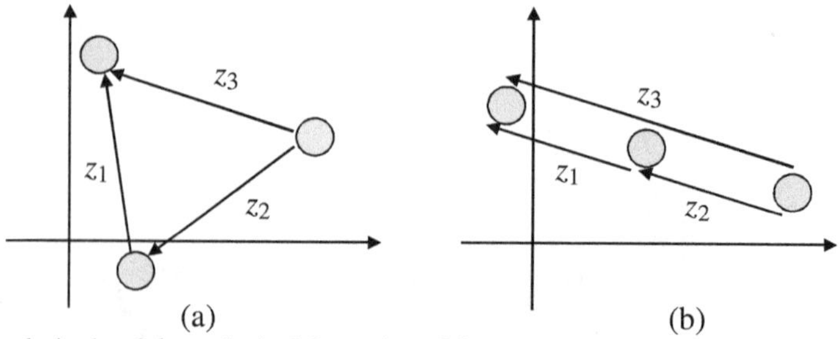

Fig. 2.8 The desired and the undesired formation of three agents with a cyclic graph: The desired formation is the equilateral triangle as (a) and the undesired formation (b) is a line.

$$D = \begin{pmatrix} 1 & 0 & 1 \\ -1 & 1 & 0 \\ 0 & -1 & -1 \end{pmatrix}. \tag{2.89}$$

The set of equilibria, given in (2.33), indicates that

$$\psi(z) = [\psi_1(z_1)^T, \psi_2(z_2)^T, \psi_3(z_3)^T]^T \in \mathcal{N}(D \otimes I_p). \tag{2.90}$$

A simple computation of the null space of D yields

$$\psi_1(z_1) = \psi_2(z_2) \tag{2.91}$$
$$\psi_1(z_1) = -\psi_3(z_3). \tag{2.92}$$

Since the undesired formation in Fig. 2.8(b) is collinear, $\frac{z_k}{|z_k|}$'s are the same. We then use (2.87) to reduce (2.91) and (2.92) to

$$\sigma_1(|z_1|) = -\sigma_3(|z_3|) \tag{2.93}$$
$$\sigma_1(|z_1|) = \sigma_2(|z_2|), \tag{2.94}$$

which have a unique solution $(|z_1| = s_1^*, |z_2| = s_2^*, |z_3| = |z_1 + z_2| = s_1^* + s_2^*)$ since $\sigma_k(\cdot)$, $k = 1,2,3$ are increasing and onto. Thus, the set of points where $|z_1| = s_1^*$, $|z_2| = s_2^*$ and $|z_3| = s_1^* + s_2^*$ constitute new equilibria as in Figure 2.8(b) and such desired cannot be eliminated with the choice of the function $\sigma_k(\cdot)$. Property 2.1 then fails in this formation control design and global stabilization of the desired formation is not possible for cyclic graphs.

For agents modeled as double integrators with uniform mass and damping, the following example proves that the undesired equilibria in Fig. 2.8(b) are unstable. In fact, this instability result can be extended to any graph that contains only one cycle. We refer interested readers to [9] for details.

Example 2.4 (Instability of the undesired formation of three agents).

Consider the undesired formation in Fig. 2.8(b). We first find out s_1^* and s_2^* from (2.93) and (2.94). We take $\sigma_k(\cdot)$, $k = 1,2,3$, as in (2.85) with $d_k = 1$. It follows from

(2.93) and (2.94) that

$$1-\frac{1}{s_1^*}=-(1-\frac{1}{s_1^*+s_2^*}) \tag{2.95}$$

and

$$s_1^*=s_2^*, \tag{2.96}$$

which yield $s_1^*=s_2^*=\frac{3}{4}$. This means that on the undesired formation, $|z_1|=|z_2|=\frac{3}{4}$ and $|z_3|=\frac{3}{2}$.

We assume that the agents have uniform mass $m_i=1$ in (2.16) and uniform damping $k_i=k>0$ in (2.17). Without loss of generality, we also let $v(t)=0$. It then follows from (2.16), (2.17), (2.26) and (2.89) that the closed-loop system for these three agents is given by

$$\ddot{x}_1+k\dot{x}_1+\psi(z_1)+\psi(z_3)=0 \tag{2.97}$$

$$\ddot{x}_2+k\dot{x}_2-\psi(z_1)+\psi(z_2)=0 \tag{2.98}$$

$$\ddot{x}_3+k\dot{x}_3-\psi(z_2)-\psi(z_3)=0 \tag{2.99}$$

where $\psi(z_k)$ is obtained from (2.87) and (2.85) as

$$\psi(z_k)=\frac{|z_k|-1}{|z_k|^2}z_k. \tag{2.100}$$

To show the instability of the undesired formation, we linearize the closed-loop system around the undesired formation $\dot{x}_i=0$, $i=1,2,3$ and $z_k=z_k^u$, $k=1,2,3$, where z_k^u denotes an undesired equilibrium of z_k. Letting $\delta z_k=z_k-z_k^u$, $k=1,2,3$, we obtain the linearized dynamics:

$$\begin{pmatrix}\ddot{x}_1\\ \ddot{x}_2\\ \ddot{x}_3\\ \dot{\delta z_1}\\ \dot{\delta z_2}\\ \dot{\delta z_3}\end{pmatrix}=\underbrace{\left(\begin{array}{ccc|ccc} -kI_p & \mathbf{0}_p & \mathbf{0}_p & -\frac{\partial\psi}{\partial z}\big|_{z_1^u} & \mathbf{0}_p & -\frac{\partial\psi}{\partial z}\big|_{z_3^u}\\ \mathbf{0}_p & -kI_p & \mathbf{0}_p & \frac{\partial\psi}{\partial z}\big|_{z_1^u} & -\frac{\partial\psi}{\partial z}\big|_{z_2^u} & \mathbf{0}_p\\ \mathbf{0}_p & \mathbf{0}_p & -kI_p & \mathbf{0}_p & \frac{\partial\psi}{\partial z}\big|_{z_2^u} & \frac{\partial\psi}{\partial z}\big|_{z_3^u}\\ \hline I_p & -I_p & \mathbf{0}_p & \mathbf{0}_p & \mathbf{0}_p & \mathbf{0}_p\\ \mathbf{0}_p & I_p & -I_p & \mathbf{0}_p & \mathbf{0}_p & \mathbf{0}_p\\ I_p & \mathbf{0}_p & -I_p & \mathbf{0}_p & \mathbf{0}_p & \mathbf{0}_p \end{array}\right)}_{\mathbf{A}}\begin{pmatrix}\dot{x}_1\\ \dot{x}_2\\ \dot{x}_3\\ \delta z_1\\ \delta z_2\\ \delta z_3\end{pmatrix} \tag{2.101}$$

$$=\left(\begin{array}{c|c}\mathbf{A}_{11} & \mathbf{A}_{12}\\ \hline \mathbf{A}_{21} & \mathbf{A}_{22}\end{array}\right)\begin{pmatrix}\dot{x}_1\\ \dot{x}_2\\ \dot{x}_3\\ \delta z_1\\ \delta z_2\\ \delta z_3\end{pmatrix} \tag{2.102}$$

where

$$\frac{\partial\psi}{\partial z}\Big|_{z^u}=\frac{|z^u|-1}{|z^u|^2}I_p+\left[-\frac{1}{|z^u|^3}+\frac{2}{|z^u|^4}\right]z^u(z^u)^T. \tag{2.103}$$

To show that the undesired formation is unstable, we only need to demonstrate that $\mathbf{A}$ has an eigenvalue with positive real part. Towards this end, we solve

$$\left(\begin{array}{c|c}\mathbf{A}_{11} & \mathbf{A}_{12}\\ \hline \mathbf{A}_{21} & \mathbf{A}_{22}\end{array}\right)\left(\frac{\mu_1}{\mu_2}\right) = \lambda\left(\frac{\mu_1}{\mu_2}\right) \tag{2.104}$$

for λ, and obtain

$$-k\mu_1 + \mathbf{A}_{12}\mu_2 = \lambda\mu_1 \tag{2.105}$$

$$\mathbf{A}_{21}\mu_1 = \lambda\mu_2. \tag{2.106}$$

Multiplying (2.105) with λ and substituting (2.106) leads to

$$\lambda^2\mu_1 + k\lambda\mu_1 - \mathbf{A}_{12}\mathbf{A}_{21}\mu_1 = 0. \tag{2.107}$$

By choosing μ_1 as the eigenvectors of $\mathbf{A}_{12}\mathbf{A}_{21}$, we obtain the eigenvalues of $\mathbf{A}$ as the solutions to the following equations

$$\lambda^2 + k\lambda - \bar{\lambda}_i = 0, \quad k > 0, \quad i = 1, \cdots, 3p \tag{2.108}$$

where $\bar{\lambda}_i$ is the ith eigenvalue of $\mathbf{A}_{12}\mathbf{A}_{21}$.

We next compute

$$\mathbf{A}_{12}\mathbf{A}_{21} = \begin{pmatrix} -\frac{\partial\psi}{\partial z}\Big|_{z_1^u} - \frac{\partial\psi}{\partial z}\Big|_{z_3^u} & \frac{\partial\psi}{\partial z}\Big|_{z_1^u} & \frac{\partial\psi}{\partial z}\Big|_{z_3^u} \\ \frac{\partial\psi}{\partial z}\Big|_{z_1^u} & -\frac{\partial\psi}{\partial z}\Big|_{z_1^u} - \frac{\partial\psi}{\partial z}\Big|_{z_2^u} & \frac{\partial\psi}{\partial z}\Big|_{z_2^u} \\ \frac{\partial\psi}{\partial z}\Big|_{z_3^u} & \frac{\partial\psi}{\partial z}\Big|_{z_2^u} & -\frac{\partial\psi}{\partial z}\Big|_{z_3^u} - \frac{\partial\psi}{\partial z}\Big|_{z_2^u} \end{pmatrix}. \tag{2.109}$$

Note from (2.103) that $\frac{\partial\psi}{\partial z}\big|_{z_k^u}$, $k = 1, 2, 3$, are symmetric. Thus, $\mathbf{A}_{12}\mathbf{A}_{21}$ is also symmetric. Then if the matrix $\mathbf{A}_{12}\mathbf{A}_{21}$ has a positive eigenvalue, there exists a positive root of (2.108) and therefore $\mathbf{A}$ is unstable.

To show that $\mathbf{A}_{12}\mathbf{A}_{21}$ has a positive eigenvalue, we recall that on the undesired formation, z_k^u's are collinear, which means that there exists a $\tilde{z} \in \mathbb{R}^p$ such that $\tilde{z} \perp z_k^u$, $\forall k$. This implies from (2.103) that

$$\frac{\partial\psi}{\partial z}\Big|_{z^u}\tilde{z} = \frac{|z^u| - 1}{|z^u|^2}\tilde{z}. \tag{2.110}$$

Choosing $\zeta = [a\ b\ c]^T \otimes \tilde{z}$, where the scalars a, b, c will be specified later, and using (2.110), we obtain

$$\mathbf{A}_{12}\mathbf{A}_{21}\zeta = \left[\begin{pmatrix} -\frac{|z_1^u|-1}{|z_1^u|^2} - \frac{|z_3^u|-1}{|z_3^u|^2} & \frac{|z_1^u|-1}{|z_1^u|^2} & \frac{|z_3^u|-1}{|z_3^u|^2} \\ \frac{|z_1^u|-1}{|z_1^u|^2} & -\frac{|z_1^u|-1}{|z_1^u|^2} - \frac{|z_2^u|-1}{|z_2^u|^2} & \frac{|z_2^u|-1}{|z_2^u|^2} \\ \frac{|z_3^u|-1}{|z_3^u|^2} & \frac{|z_2^u|-1}{|z_2^u|^2} & -\frac{|z_2^u|-1}{|z_2^u|^2} - \frac{|z_3^u|-1}{|z_3^u|^2} \end{pmatrix}\begin{pmatrix} a \\ b \\ c \end{pmatrix}\right] \otimes \tilde{z}. \tag{2.111}$$

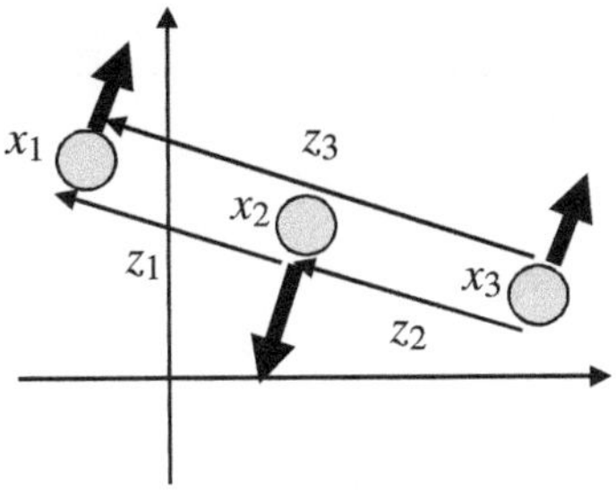

Fig. 2.9 The unstable eigenvector of the undesired formation corresponds to motion in the direction indicated by the bold arrows.

We now substitute $|z_1^u| = |z_2^u| = \frac{3}{4}$ and $|z_3^u| = \frac{3}{2}$ into (2.111) and get

$$\mathbf{A}_{12}\mathbf{A}_{21}\zeta = \left[\underbrace{\begin{pmatrix} \frac{2}{9} & -\frac{4}{9} & \frac{2}{9} \\ -\frac{4}{9} & \frac{8}{9} & -\frac{4}{9} \\ \frac{2}{9} & -\frac{4}{9} & \frac{2}{9} \end{pmatrix}}_{\mathbf{B}} \begin{pmatrix} a \\ b \\ c \end{pmatrix} \right] \otimes \tilde{z}. \tag{2.112}$$

The matrix $\mathbf{B}$ has a positive eigenvalue $\bar{\lambda} = \frac{4}{3}$ associated with an eigenvector $[-1\ 2\ -1]^T$. By choosing $[a\ b\ c]^T = [-1\ 2\ -1]^T$, we rewrite (2.112) as

$$\mathbf{A}_{12}\mathbf{A}_{21}\zeta = \bar{\lambda}\zeta, \tag{2.113}$$

which shows that $\mathbf{A}_{12}\mathbf{A}_{21}$ has a positive eigenvalue $\bar{\lambda}$. Therefore, the undesired formation in 2.8(b) is unstable.

The unstable eigenvector $[-1\ 2\ -1]^T \otimes \tilde{z}$ corresponds to motion in the direction shown by the bold arrows in Fig. 2.9. We interpret the unstable growth in this direction by returning to the mass-spring analogy. Since $|z_1^u| = |z_2^u| < 1$ and $|z_3^u| > 1$, springs 1 and 2 are squeezed while spring 3 is stretched. The motion in Fig. 2.9 increases $|z_1|$ and $|z_2|$ towards their natural length of one. □

Because the undesired formation in Fig. 2.8(b) is unstable, we conclude generic convergence to the desired formation in Fig. 2.8(a) from all initial conditions except for those that lie on the stable manifolds of the unstable equilibria. The numerical example in Fig. 2.10 shows the convergence to the desired formation with the design (2.16) and (2.17) for three agents. In this example, the reference velocity $v(t)$ is chosen as $[0.1\ 0.1]^T$.

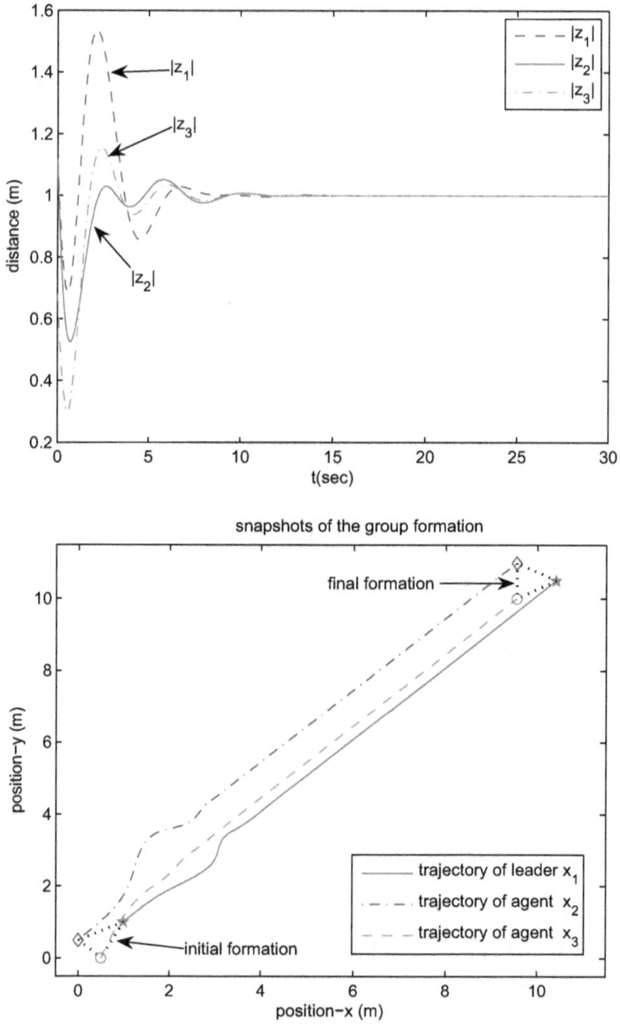

Fig. 2.10 Snapshots of the formation for the passivity-based design (2.16) and (2.17): The three relative positions, z_1, z_2 and z_3 denote $x_1 - x_2$, $x_3 - x_1$ and $x_2 - x_3$. The agents x_1, x_2 and x_3 are represented by $\star$, $\diamond$ and $\circ$, respectively.

2.7.2 Existence and Uniqueness of a Formation Shape

A key consideration in the distance-based formation control problem is whether a given set of desired relative distances d_k's, even admits an equilibrium in the closed loop, and, if so, whether the equilibrium is unique.

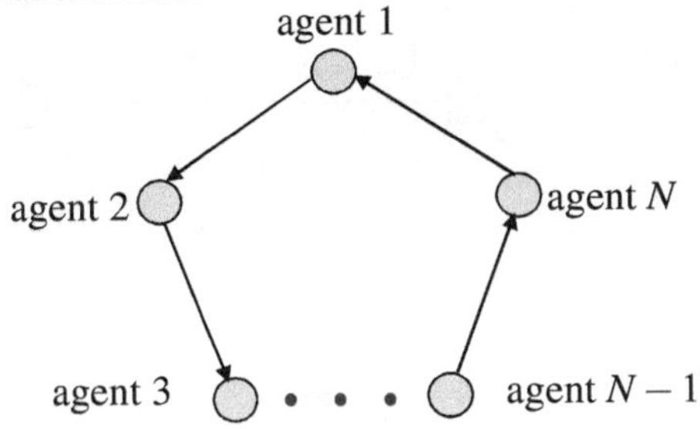

Fig. 2.11 The ring graph of N agents. The directions of the N links are assigned such that the positive end of each link is the negative end of the next link in the sequence.

Given d_k's, a desired formation equilibrium exists if the constraint (2.7) is satisfied. If the graph is acyclic, (2.7) holds for any $d_k > 0$, $k = 1, \cdots, \ell$. So we only need to consider the cyclic graph case.

As an example, consider a group of N agents that form a ring graph, i.e., each agent has exactly two neighbors. A ring graph has only one cycle and the numbers of links and nodes are the same, i.e., $N = \ell$. As shown in Fig. 2.11, we assign the orientation of the ring graph such that the positive end of each link is the negative end of the next link in the sequence. We define $z_i = x_{i+1} - x_i$, $\forall i = 1, \cdots, N-1$ and $z_N = x_1 - x_N$, and obtain

$$z_j = - \sum_{k=1, k \neq j}^{N} z_k, \quad \forall j. \tag{2.114}$$

This equality must be satisfied at the desired formation. Therefore, we obtain

$$|z_j| = |\sum_{k=1, k \neq j}^{N} z_k| \leq \sum_{k=1, k \neq j}^{N} |z_k|, \quad \forall j \tag{2.115}$$

$$\Rightarrow d_j \leq \sum_{k=1, k \neq j}^{N} d_k, \quad \forall j. \tag{2.116}$$

When $N = 3$ and the desired formation is a triangle, (2.116) reduces to the triangle inequality, that is, the sum of the lengths of any two sides of the triangle must be greater than the length of the other side. Thus, the choice of d_k is constrained by (2.116). If the graph contains multiple cycles, multiple constraints similar to (2.116) must be satisfied for d_k's so that a desired formation exists.

Once we establish that a desired formation exists for a given set of d_k's, the shape of the desired formation may not be unique if we do not specify enough number of desired relative distances. We illustrate this using a four-agent example below. More formal analysis on the uniqueness of a formation shape using "rigidity" can be found in [101, 46, 72, 43].

Example 2.5 (Stabilizing a rectangle formation of four agents).

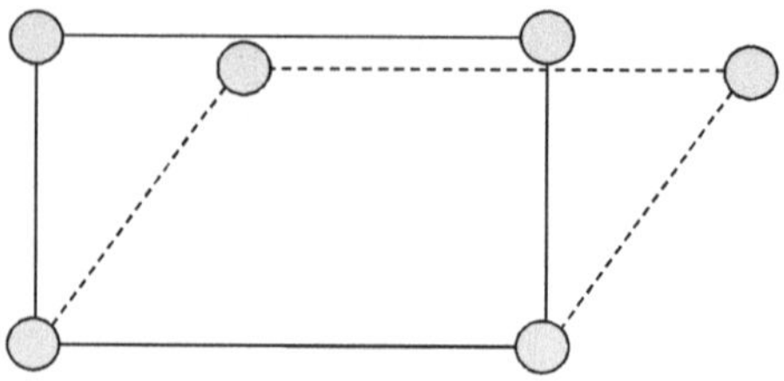

Fig. 2.12 The square formation can collapse to a parallelogram or eventually to a line.

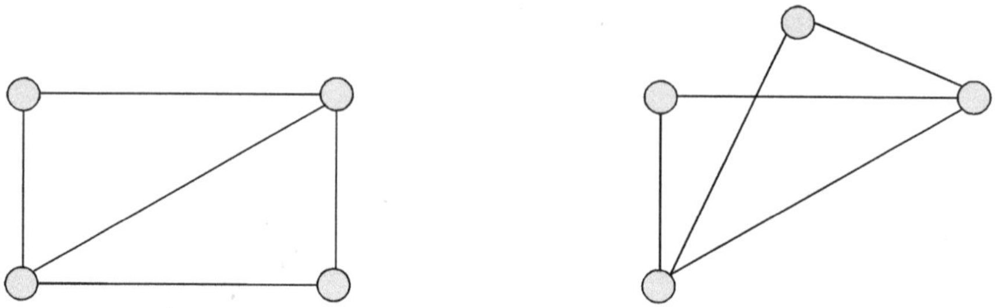

(a) (b)

Fig. 2.13 Two possible desired formations when one diagonal link is specified.

Suppose that we want to stabilize four agents to a rectangle formation (solid lines in Fig. 2.12). Initially, we only specify the desired relative distances of the four sides. As illustrated in Fig. 2.12, the agents may converge to a rectangle or to a parallelogram or even to a line since all those shapes are in the target set (2.50). In fact, there exist infinitely many formations (up to rigid translation and rotation) in (2.50). Thus, specifying the lengths of four sides is not enough to guarantee the desired rectangle formation (up to rigid translation and rotation).

We then add a diagonal link and specify its length. Then there exist only two possible formations (up to rigid translation and rotation) in Fig. 2.13. Thus, if the agents converge to the target set (2.50), they will converge to either of the two shapes in Fig. 2.13.

If we also specify the length of the other diagonal link as shown in Fig. 2.14, we eliminate the existence of the formation in Fig. 2.13(b). In this case, if the agents converge to the desired equilibria, they converge to the desired rectangle formation. □

2.8 Distance-based or Position-based?

We have seen two types of formation control, both of which can be designed with the passivity-based framework. We now present a comparison between these two

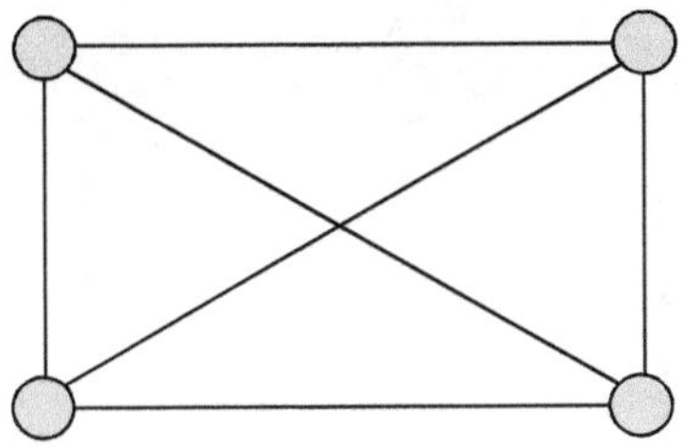

Fig. 2.14 If we specify the desired lengths of all the links, the desired formation is unique.

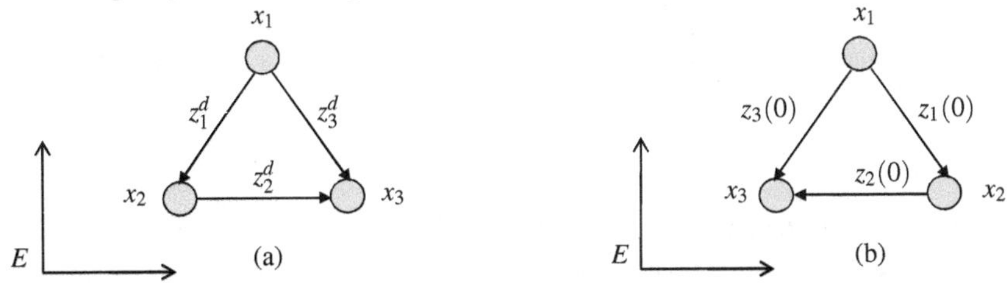

Fig. 2.15 (a) Desired formation in terms of z_k^d, $k = 1,2,3$. (b) Initial formation of the three agents. Even if the initial formation (b) is an equilateral triangle, agents 2 and 3 will still swap their positions to match the desired formation in (a).

formulations and illustrate the situations under which one formulation is preferable to the other.

• *Equilibria.*

For the distance-based formation control, the desired equilibria in (2.50) are spheres while for the position-based formation control, the desired equilibrium in (2.51) is a single point. The difference in these equilibria sets reflects different control objectives. When the agents need to maintain specific bearings and distances with respect to their neighbors, the position-based formation control is more suitable. In the case where cooperative tasks only require the shape of the formation rather than a specific orientation of the formation, the distance-based formation control is preferable. This is because the position-based formation control may put stringent constraints on relative positions and sacrifice the flexibility of the group motion, as we illustrate below.

Example 2.6. Consider a group of three agents $x_i \in \mathbb{R}^2$, $i = 1,2,3$. Suppose that the desired formation is an equilateral triangle with side length 1. One way to achieve this desired formation is to apply the position-based formation control by specifying desired relative positions between agents. We let $z_1 = x_2 - x_1$, $z_2 = x_3 - x_2$ and $z_3 = x_3 - x_1$, and choose $z_1^d = [-\frac{1}{2} \; -\frac{\sqrt{3}}{2}]$, $z_2^d = [1 \; 0]^T$ and $z_3^d = [\frac{1}{2} \; -\frac{\sqrt{3}}{2}]^T$ in the frame of E, as in Figure 2.15(a).

Let the initial formation of the three agents shown in Figure 2.15(b) be $z_1(0) = z_3^d$, $z_2(0) = -z_2^d$ and $z_3(0) = z_1^d$, which means that it is already an equilateral triangle with side length 1. However, this equilateral formation does not match the desired targets (2.51). Since (2.51) is globally attractive by the position-based formation

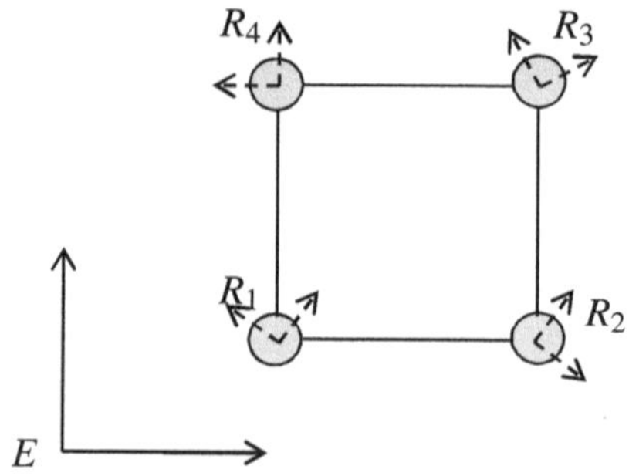

Fig. 2.16 Four planar agents in a global frame E: R_i represents a local frame for agent i. The desired formation is shown as an equilateral square.

control design, the agents will start moving away from the initial formation in Figure 2.15(b) towards the desired formation in Figure 2.15(a), which results in unnecessary time and control energy consumption. □

• *Control Design and Stability.*

The difference in equilibria leads to different control designs: The design in Section 2.7.1 employs nonlinear potential functions to achieve the distance-based formation control while position-based formation control can be realized by linear feedback laws, such as the design in (2.74).

Moreover, because of the difference in equilibria sets, the distance-based formation control stabilizes the desired formation only locally when the graph contains cycles while the position-based formation control is able to globally stabilize the desired formation. This is because Property 2.1 is satisfied for the position-based formation control but not for the distance-based formation control. Because the position-based formation control is globally stabilizable, it has been applied to different cooperative control problems, including formation control of unicycles [83, 41].

• *Desired Formation Specification.*

The difference in the equilibria sets (2.50) and (2.51) also results in different specifications of the desired formation. For example, consider four planar agents in Fig. 2.16. The coordinate E is a global frame while R_i, $i = 1, \cdots, 4$, is agent i's local frame, possibly different from E. The desired formation is an equilateral square shown in Fig. 2.16. To specify a desired formation using (2.50), one only needs to determine the desired relative distances, which is invariant in different frames. This implies that the desired distances can be specified in either E or R_i's. However, for position-based formation control, the desired relative position z^d must be prescribed in one common frame, such as E in Fig. 2.16.

One subtlety in specifying a desired formation is how to guarantee a *unique* desired formation. For position-based formation control, specifying desired relative positions of $N-1$ links is sufficient if these $N-1$ links can form a connected graph. This is because once the desired relative positions of these $N-1$ links are fixed, the desired relative positions between any two agents are also fixed. As an illustration, consider Fig. 2.16 and suppose that we specify the relative positions for the links between agents 1 and 2, between agents 2 and 3, and between agents 3 and 4. These three links form a connected graph and thus all the other relative positions, such as relative positions between agents 4 and 1, between agents 2 and 4, are uniquely determined. However, for distance-based formation control, specifying the desired lengths of $N-1$ links may not enough to ensure a unique formation, as we already illustrated in Example 2.5.

- *Implementation of Control Laws.*

In practice, the relative position z_k's are obtained in each agent's local frame. If precise global frame information is available, the agents may transform the local measurements of z_k to the global coordinates for implementation. However, in some applications, such as space interferometry sensing, the global frame information may be imprecise or unavailable. In this case, we show that distance-based formation control can be easily implemented without knowledge of any global frame information while position-based formation control requires the knowledge of a common frame in which the desired relative positions z^d are specified.

For illustration, we assume double integrator dynamics for the agents and rewrite the control laws for distance-based and position-based formation control as

$$\text{position-based:}\quad \ddot{x}_i = \tau_i = -k_i\dot{x}_i - \sum_{i=1}^{\ell} d_{ik}(z_k - z_k^d) \tag{2.117}$$

$$\text{distance-based:}\quad \ddot{x}_i = \tau_i = -k_i\dot{x}_i - \sum_{i=1}^{\ell} d_{ik}\log(\frac{|z_k|}{d_k})\frac{1}{|z_k|}z_k \tag{2.118}$$

where we take $v(t) = 0_p$. Suppose that (2.117)-(2.118) are written in a global frame E. Then z_k^d's must be specified in E. When E is not available, each agent implements ${}^i\tau_i$, the τ_i vector represented in agent i's frame[2]. Let R_i be the agent i's frame represented in E. Then ${}^i\tau_i$'s are given by

$$\text{position-based:}\quad {}^i\tau_i = -k_iR_i^T\dot{x}_i - \sum_{i=1}^{\ell} d_{ik}(R_i^Tz_k - R_i^Tz_k^d) \tag{2.119}$$

$$\text{distance-based:}\quad {}^i\tau_i = -k_iR_i^T\dot{x}_i - \sum_{i=1}^{\ell} d_{ik}\log(\frac{|z_k|}{d_k})\frac{1}{|z_k|}R_i^Tz_k. \tag{2.120}$$

It then becomes evident that both (2.119) and (2.120) require agent i's velocity represented in R_i (i.e., the term $R_i^T\dot{x}_i$) and the relative position z_k represented in R_i (i.e., the term $R_i^Tz_k$). In addition, the position-based formation control also needs $R_i^Tz_k^d$, which cannot be computed if the global frame E is not known. Thus, we conclude that distance-based formation control is more suitable for applications with no global frame information.

---

[2] For vector representations in different frames, we refer readers to Appendix B.12.1.

2.9 Summary

In this chapter, we employed passivity as a design tool for a class of group coordination problems where the topology of information exchange between agents is bidirectional. We exploited the symmetry inherent in the undirected graph and represented it as a passivity-preserving structure (pre-multiplication of $D^T \otimes I_p$ and post-multiplication by its transpose as in Fig. 2.2). We used this structure to develop a passivity-based design framework that yields a broad class of decentralized and scalable cooperative control laws for complex and heterogeneous agent dynamics. In addition to stabilizing feedback rules, the passivity-based design framework constructs a Lur'e-type Lyapunov function. As we will illustrate in Chapters 3, 4 and 6, this Lyapunov function serves as a starting point for several adaptive designs that enhance robustness of group motion.

We next applied the passivity-based design framework to agreement problems. We developed a class of decentralized protocols that achieve agreement of agents. We also studied the position-based and the distance-based formation control. For the position-based formation control, we showed that it can be transformed to an agreement problem, which means that the desired formation is guaranteed. In the distance-based formation control, we showed that the desired formation is only locally asymptotically stable for cyclic graphs because Property 2.1 fails. We then proved the instability of the undesired formations for a three-agent example and concluded generic convergence to the desired formation. We also discussed how to specify a unique and feasible formation shape. Finally, a comparison between the position-based and the distance-based formation control was presented.

2.10 Notes and Related Literature

- The use of Schur decomposition in Example 2.3 follows [47, Theorem 3].
- The passivity-based framework in this chapter was developed in [5].
- Related Literature on agreement and formation control: A rapidly-growing literature has been witnessed in the field of agreement. See e.g., [102, 109] for a summary. Applications of formation control can be found in the survey papers [117, 118, 26, 93]. Reference [98] first applied potential function method to the formation control with undirected information topology. A flocking algorithm was studied in [131] under time-varying communication graphs. In [47], the formation of multiple vehicles with linear identical dynamics was investigated. Based on a decentralized simultaneous estimation and control framework, the authors in [150] studied formation control using geometric moments. In [44], the position-based formation control was formulated as an optimization problem and a distributed receding horizon controller was proposed. Reference [123] considered optimal formation control of linear agent dynamics by using relative position and communicating estimates. In [124], a parallel estimator was developed for controlling formation of linear dynamical agents with directed graphs. Reference [152] employed Jacobi shape theory

to decouple translational formation dynamics from shape and orientation dynamics. The proposed cooperative control laws locally stabilize the desired formation shape. Formation control with directed graphs has also been investigated in [22, 23, 55, 4].

• For directed graphs, significant results have been obtained using a number of different approaches, such as the use of Laplacian properties for directed graphs in [103, 47, 109, 79], input-to-state stability [135], passive decomposition of group dynamics [76], eigenvalue structure of circulant matrices [87], set-valued Lyapunov theory in [90], and contraction analysis [31, 32]. In particular, recent research in [29, 28] also employed passivity as a tool for agreement of nonlinear systems. The results [29, 28] are applicable to strongly connected directed graphs for relative degree one agents. The passivity-based framework in this book allows agent dynamics to be relative degree higher than one for undirected graphs.

• Step 1 in the passivity-based framework may not be applicable to certain classes of dynamical systems, such as nonholonomic agents. Significant research has been conducted when the agents are modeled as unicycles. In [77], the authors considered a group of unit speed unicycles and proposed designs to achieve different group formations. A leader-following approach was introduced in [48] to ensure a desired group formation, where each unicycle maintains desired relative bearings and distances with respect to its neighbors. The control algorithms were based input-output linearization. Reference [88] studied cooperative formation control of multiple unicycles by assigning each agent a desired trajectory to track. The tracking errors decrease as feedback gains increase. In [83], formation control of unicycles was studied in the position-based formulation and necessary and sufficient graphical conditions were obtained. Reference [87] employed eigenvalue structure of circulant matrices in cyclic pursuit formation. Agreement of positions and orientations of unicycles was considered in [37] and discontinuous time-invariant control laws were analyzed using nonsmooth analysis. For dynamical nonholonomic agents, backstepping is a useful tool to transform coordination laws from kinematic level to dynamic level [42, 39]. In [50], formation control with general agent dynamics was formulated as a nonlinear output regulation problem.

Chapter 3
Adaptive Design for Reference Velocity Recovery: Internal Model Approach

3.1 Introduction

The passivity-based design in Chapter 2 assumed that the reference velocity of the group is available to each agent and developed control laws that made use of this information. A more realistic situation is when a leader, say agent 1, in the group possesses this information. In this chapter, we exploit the design flexibility offered by the passivity-based framework and develop adaptive designs with which the other agents reconstruct reference velocity information. We first illustrate with an example that if the agents do not have the same reference velocity information, objectives A1 and A2 cannot be guaranteed. Assuming that the reference velocity is constant or periodically time-varying, we then propose a *basic* adaptive design that allows the other agents estimate the reference velocity. This basic adaptive design preserves the passivity properties proven in Theorem 2.1 and recovers objective A2. The derivation of this adaptive design follows techniques from the regulation problem studied in [21, 19, 64] and is referred to as the internal model approach because this design contains a model of the dynamic structure of the reference velocity.

With the basic adaptive design, we next present an example which shows that the estimates of the reference velocity may not converge to their true value when the reference velocity is time-varying. This means that tracking of the reference velocity (objective A1) is not guaranteed. However, the basic adaptive design guarantees tracking of the reference velocity in several special cases, such as, when the reference velocity is constant, or when the desired target set is the origin.

In the situation where tracking of reference velocity fails for the basic adaptive design, we propose an *augmented* adaptive design to guarantee tracking of the reference velocity without the restrictions discussed above. The main idea in the augmented design is to ensure that the relative velocities between agents converge to zero, thereby guaranteeing that all agents converge to the reference velocity. By including the relative velocity feedback in the redesign, we recover the stability result of the basic adaptive design while achieving tracking of the reference velocity.

3.2 Why Adaptation?

The design (2.8), (2.11) and (2.23) assumes that the reference velocity $v(t)$ is available to each agent. In practice, the reference velocity information may be available only to a leader, say, the first agent, while the other agents may have incorrect $v(t)$ information. We then let $v_i(t)$ be the reference velocity of agent i, $i = 1, \cdots, N$, where $v_1(t) = v(t)$ and $v_i(t) \neq v(t)$, $\forall i \neq 1$, which means that agents $i = 2, \cdots, N$, have incorrect reference velocity information while agent 1 has the correct one. We now consider the example in (2.69) and examine the performance of the group motion due to different $v_i(t)$'s.

For simplicity, we assume that $m_i = 1$, $P_k(z_k)$ satisfies (2.71) with $\delta_k = 1$, and $v_i(t)$'s are constant. Using (2.69), we obtain the closed-loop system

$$\ddot{x} = -k(\dot{x} - \bar{v}) - (D \otimes I_p)(z - z^d) \tag{3.1}$$

where $\bar{v} = [v_1^T, \cdots, v_N^T]^T$. Denoting

$$\tilde{v}_i = v_i - \frac{1}{N} \sum_{i=1}^{N} v_i \tag{3.2}$$

and $\tilde{v} = [\tilde{v}_1^T, \cdots, \tilde{v}_N^T]^T$, we obtain from (3.1) that

$$\ddot{x} = -k \left(\dot{x} - 1_N \otimes \sum_{i=1}^{N} \frac{1}{N} v_i \right) - (D \otimes I_p)(z - z^d) + k\tilde{v}. \tag{3.3}$$

Using the transformation

$$\mathbf{x}(t) := x(t) - x_c - \frac{1}{N} \left(1_N \otimes \sum_{i=1}^{N} v_i t \right), \tag{3.4}$$

where x_c is as in (2.52), and recalling (2.54), we rewrite (3.3) as

$$\ddot{\mathbf{x}} + k\dot{\mathbf{x}} + (L \otimes I_p)\mathbf{x} = k\tilde{v}. \tag{3.5}$$

According to the result in Section 2.6.1, if $k\tilde{v} = 0$, system (3.5) has a globally asymptotically stable equilibrium at $\dot{\mathbf{x}} = 0$ and $\mathbf{z} = 0$. Thus, nonzero $k\tilde{v}$ can be considered as a constant disturbance to (3.5). This constant disturbance shifts the equilibrium of (3.5). To find out this shifted equilibrium, we rewrite (3.5) as

$$\begin{pmatrix} \ddot{\mathbf{x}} \\ \dot{\mathbf{z}} \end{pmatrix} = \left[\begin{pmatrix} -kI_N & -D \\ D^T & \mathbf{0}_\ell \end{pmatrix} \otimes I_p \right] \begin{pmatrix} \dot{\mathbf{x}} \\ \mathbf{z} \end{pmatrix} + \begin{pmatrix} k\tilde{v} \\ 0 \end{pmatrix}, \tag{3.6}$$

and compute the equilibrium of (3.6) by solving

$$\left[\begin{pmatrix} -kI_N & -D \\ D^T & \mathbf{0}_\ell \end{pmatrix} \otimes I_p \right] \begin{pmatrix} \dot{\mathbf{x}} \\ \mathbf{z} \end{pmatrix} = - \begin{pmatrix} k\tilde{v} \\ 0 \end{pmatrix} \tag{3.7}$$

which leads to

$$(D^T \otimes I_p)\dot{\mathbf{x}} = 0 \tag{3.8}$$

and

$$-k\dot{\mathbf{x}} + (D \otimes I_p)\mathbf{z} = -k\tilde{\mathbf{v}}. \tag{3.9}$$

Using the fact that 1_N spans $\mathcal{N}(D^T)$ and noting from (3.2) that

$$(1_N \otimes I_p)^T(-k\tilde{\mathbf{v}}) = 0, \tag{3.10}$$

we obtain from (3.9) that

$$(1_N \otimes I_p)^T(-k\dot{\mathbf{x}} + (D \otimes I_p)\mathbf{z}) = -k(1_N \otimes I_p)^T\dot{\mathbf{x}} = (1_N \otimes I_p)^T(-k\tilde{\mathbf{v}}) = 0. \tag{3.11}$$

Because (3.8) implies that $\dot{\mathbf{x}} = 1_N \otimes c$ for some $c \in \mathbb{R}^p$, it follows from (3.9) and (3.11) that

$$\dot{\mathbf{x}} = 0 \tag{3.12}$$

and

$$(D \otimes I_p)\mathbf{z} = -k\tilde{\mathbf{v}}. \tag{3.13}$$

Since (3.10) implies that $-k\tilde{\mathbf{v}} \perp \mathcal{N}(D^T \otimes I_p)$, $-k\tilde{\mathbf{v}}$ lies in $\mathcal{R}(D \otimes I_p)$, which means that there exists a nonzero $\bar{\mathbf{z}}$ satisfying $(D \otimes I_p)\bar{\mathbf{z}} = -k\tilde{\mathbf{v}}$. Then the equilibrium of (3.6) is given by

$$\dot{\mathbf{x}} = 0 \quad \text{and} \quad \mathbf{z} = \bar{\mathbf{z}}. \tag{3.14}$$

We note from (3.4) that $\dot{\mathbf{x}} = 0$ means that the velocities of all the agents converge to $\frac{1}{N}\sum_{i=1}^N v_i$ rather than the correct reference velocity v. The steady state of $\mathbf{z}$ is also shifted to $\bar{\mathbf{z}} \neq 0$, which implies that the desired target $z = z^d$ cannot be reached.

From this simple example, we see that if the agents do not share the same reference velocity information, steady state errors arise in both the reference velocity tracking and the convergence to the desired target sets. In the following sections, we will propose adaptive designs with which the agents recover the reference velocity information and eliminate the steady state errors. In particular, we will present two approaches: the Internal Model approach and the Parameterization approach. The internal model approach assumes the reference velocity to be constant or periodically time-varying while the parameterization approach developed in Chapter 4 parameterizes the reference velocity as a linear combination of known time-varying basis functions with unknown constant coefficients.

3.3 Internal Model Approach: The *Basic* Design

Let $v^n(t) \in \mathbb{R}$ be the nth element of the reference velocity $v(t) \in \mathbb{R}^p$, $n = 1, \cdots, p$. In this section, we consider the situation where $v^n(t)$, $n = 1, \cdots, p$, are generated by the following exogenous systems:

$$\dot{\eta}^n = \bar{A}\eta^n \tag{3.15}$$
$$v^n(t) = H^n\eta^n, \tag{3.16}$$

in which $\bar{A} \in \mathbb{R}^{q\times q}$ satisfies $\bar{A} = -\bar{A}^T$, $\eta^n \in \mathbb{R}^q$, $\eta^n(0)$ is arbitrary and $H^n \in \mathbb{R}^{1\times q}$. The skew symmetry of $\bar{A}$ implies that the eigenvalues of $\bar{A}$ lie on the imaginary axis and thus $v(t)$ can be constant or periodically time-varying. We further make the following assumptions:

Assumption 1 *The matrix $\bar{A}$ is available to all the agents.* □

Assumption 1 is reasonable in the situation where the general dynamic structure of the reference velocity is known, but the observation matrix H^n and the initial conditions of $\eta^n(0)$ may be chosen by the leader. This is relevant, for example, when the leader makes decisions autonomously in real time.

Assumption 2 *$(\bar{A}, H^n)$ is observable, $n = 1, \cdots, p$.* □

Assumption 2 is standard. Denoting

$$\mathscr{Q}_{H^n} = \begin{pmatrix} H^n \\ H^n\bar{A} \\ \vdots \\ H^n\bar{A}^{q-1} \end{pmatrix} \in \mathbb{R}^{q\times q}, \quad n = 1, \cdots, p \tag{3.17}$$

and

$$\mathscr{Q}_H = \mathrm{diag}\left\{\mathscr{Q}_{H^1}, \cdots, \mathscr{Q}_{H^p}\right\}, \tag{3.18}$$

we note that Assumption 2 is equivalent to $\mathscr{Q}_H$ being full rank.

For compactness, we introduce

$$\eta = [(\eta^1)^T, \cdots, (\eta^p)^T]^T \in \mathbb{R}^{pq} \tag{3.19}$$

and

$$A = (I_p \otimes \bar{A}) \in \mathbb{R}^{pq\times pq}, \quad H = \mathrm{diag}\left\{H^1, \cdots, H^p\right\} \in \mathbb{R}^{p\times pq}, \tag{3.20}$$

and rewrite $v(t)$ from (3.15) and (3.16) as

$$\dot{\eta} = A\eta \tag{3.21}$$
$$v(t) = H\eta. \tag{3.22}$$

The first step in the basic adaptive design is to modify the internal feedback loop to assign an estimate of $v(t)$ to agent i. To this end, we modify (2.8) as

$$\dot{x}_1 = y_1 + v(t) \tag{3.23}$$
$$\dot{x}_i = y_i + \hat{v}_i(t) \quad i = 2, \cdots, N, \tag{3.24}$$

where $\hat{v}_i$ is agent i's estimate of $v(t)$. As demonstrated in Section 3.4, such a modification can be easily achieved based on the design in Step 1 in Chapter 2.

We next make use of the feedback (2.23) and design an update law for $\hat{v}_i$. Let $\hat{v}_i^n$ and u_i^n, $n = 1, \cdots, p$, be the nth element of $\hat{v}_i$ and u_i, respectively. For agent i, $i = 2, \cdots, N$, we design

$$\dot{\varpi}_i^n = \bar{A}\varpi_i^n + (B_i^n)^T u_i^n \tag{3.25}$$

$$\hat{v}_i^n = B_i^n \varpi_i^n, \quad n = 1, \cdots, q, \tag{3.26}$$

where $\varpi_i^n \in \mathbb{R}^q$, $\varpi_i^n(0)$ is arbitrary, and $B_i^n \in \mathbb{R}^{1\times q}$. Note that the design (3.25)-(3.26) contains the model information $\bar{A}$ of the reference velocity. The main restriction of this design is that $(\bar{A}, B_i^n)$ be observable, that is,

$$\mathcal{Q}_{B_i^n} = \begin{pmatrix} B_i^n \\ B_i^n \bar{A} \\ \vdots \\ B_i^n \bar{A}^{q-1} \end{pmatrix} \in \mathbb{R}^{q\times q} \quad \text{be full rank.} \tag{3.27}$$

Letting

$$\varpi_i = [(\varpi_i^1)^T, \cdots, (\varpi_i^p)^T]^T \in \mathbb{R}^{pq} \quad \text{and} \quad B_i = \text{diag}\left\{B_i^1, \cdots, B_i^p\right\} \in \mathbb{R}^{p\times pq}, \tag{3.28}$$

we obtain the compact form of (3.25) and (3.26) as

$$\dot{\varpi}_i = A\varpi_i + B_i^T u_i \tag{3.29}$$

$$\hat{v}_i = B_i \varpi_i. \tag{3.30}$$

The block diagram of (3.29)-(3.30) is shown in Fig. 3.1, which resembles the passivity-preserving Structure 1 in Section 1.5. In fact, because $A = -A^T$, using the storage function $\frac{1}{2}\varpi_i^T\varpi_i$, we prove the passivity from $B_i^T u_i$ to ϖ_i in Fig. 3.1. According to Structure 1, we obtain the following passivity result:

Lemma 3.1. *The system in (3.29)-(3.30) is passive from u_i to $\hat{v}_i(t)$.* □

Lemma 3.1 is essential in establishing the closed-loop stability of the basic adaptive design in Theorem 3.1.

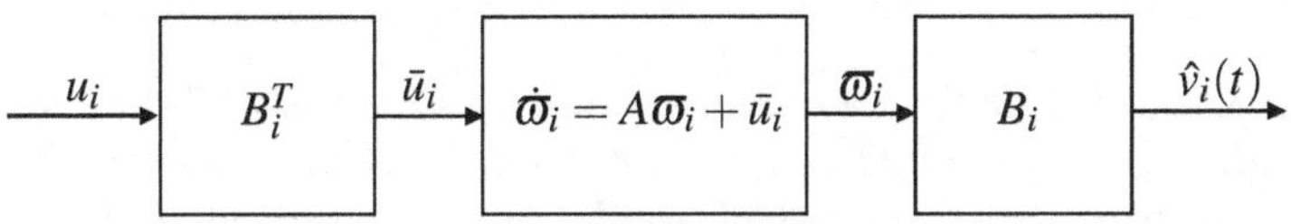

Fig. 3.1 The blockdiagram of (3.29)-(3.30) resembles the passivity-preserving Structure 1 in Section 1.5. The passivity from u_i to $\hat{v}_i$ can be established by using the storage function $\frac{1}{2}\varpi_i^T\varpi_i$.

The goal of the adaptive design is to recover the convergence results $|\dot{x}_i - v(t)| \to 0$ and $u = -(D\otimes I_p)\psi(z) \to 0$ in Theorem 2.1. Note from (3.24) that $|\dot{x}_i - v(t)| \to 0$, $\forall i = 2, \cdots, N$, is ensured by $y_i \to 0$ and $|\hat{v}_i(t) - v(t)| \to 0$. Lemma 3.2 below employs the observability conditions of $(\bar{A}, H^n)$ and $(\bar{A}, B_i^n)$ and proves that when $u \equiv 0$, (3.29)-(3.30) has a unique solution $\bar{\varpi}_i(t)$ such that $\hat{v}_i(t)$ in (3.30) is equal to $v(t)$ in (3.22). This allows us to transform the convergence $|\hat{v}_i(t) - v(t)| \to 0$ to $|\varpi_i - \bar{\varpi}_i| \to 0$.

Lemma 3.2. *Define*

$$\Sigma_i^n = \mathscr{Q}_{B_i^n}^{-1} \mathscr{Q}_{H^n}, \quad n = 1, \cdots, p, \quad i = 1, \cdots, N, \tag{3.31}$$

where $\mathscr{Q}_{H^n}$ is in (3.17) and $\mathscr{Q}_{B_i^n}$ is in (3.27). Let

$$\Sigma_i = \operatorname{diag}\left\{\Sigma_i^1, \cdots, \Sigma_i^p\right\}, \tag{3.32}$$

and

$$\bar{\varpi}_i(t) = \Sigma_i \eta(t) \tag{3.33}$$

where $\eta(t)$ is as in (3.21). Then, when $u_i \equiv 0$, $\bar{\varpi}_i(t)$ is the unique solution to (3.29)-(3.30) such that $\hat{v}_i(t) = v(t)$, where $v(t)$ is as in (3.22). □

We next define

$$\tilde{\varpi}(t) = [(\varpi_2 - \bar{\varpi}_2)^T, \cdots, (\varpi_N - \bar{\varpi}_N)^T]^T \tag{3.34}$$

where $\bar{\varpi}_i$ is in (3.33), and note that the equilibria set of (3.21)-(3.24), (3.29) and (3.30) translates to

$$\mathscr{E}^* = \{(z, \xi, \tilde{\varpi}) | \ (\xi, \ z) \in \mathscr{E} \text{ and } \tilde{\varpi} = 0\} \tag{3.35}$$

where $\mathscr{E}$ is as in (2.33). This equilibria set includes the desired equilibria set

$$\mathscr{A}^* = \{(z, \xi, \tilde{\varpi}) | \ (\xi, \ z) \in \mathscr{A} \text{ and } \tilde{\varpi} = 0\}, \tag{3.36}$$

where $\mathscr{A}$ is in (2.36). Theorem 3.1 below makes use of the passivity property in Lemma 3.1 and proves the stability of $\mathscr{A}^*$. It also shows that the trajectories $(\xi(t), z(t))$ converge to $\mathscr{E}$. Whether the convergence $\tilde{\varpi} \to 0$ is achieved or not depends on the properties of the reference velocity, the desired target sets $\mathscr{A}_k$ and the potential functions $P_k(z_k)$. We will elaborate on this after Theorem 3.1.

Theorem 3.1. *Consider the coordination laws in (3.23) and (3.24) where $v(t)$ and $\hat{v}_i(t)$ are given by (3.21)-(3.22) and (3.29)-(3.30), in which A is in (3.20) and $\bar{A} = -\bar{A}^T$, u_i is defined in (2.23) in which ψ_k, $k = 1, \cdots, \ell$, are designed as in (2.27)-(2.31), and y_i is the output of $\mathscr{H}_i$ with the input u_i, $i = 1, \cdots, N$, in which the passive block $\mathscr{H}_i$ is designed as in (2.11)-(2.15). Suppose that (3.27) holds. Then,*

i) The desired equilibria set $\mathscr{A}^*$ in (3.36) is stable;

ii) All trajectories $(z(t), \xi(t), \tilde{\varpi}(t))$ starting in $\mathscr{G} \times \mathbb{R}^{pq(N-1)}$ are bounded and converge to $\mathscr{E} \times \mathbb{R}^{pq(N-1)}$, where $\mathscr{E}$ is as in (2.33);

iii) If Property 2.1 holds, all trajectories $(z(t), \xi(t), \tilde{\varpi}(t))$ starting in $\mathscr{G} \times \mathbb{R}^{pq(N-1)}$ converge to $\mathscr{A} \times \mathbb{R}^{pq(N-1)}$. □

Proof (Lemma 3.2). When $u_i \equiv 0$, (3.29)-(3.30) reduces to

$$\dot{\varpi}_i = A\varpi_i \tag{3.37}$$
$$\hat{v}_i(t) = B_i\varpi_i(t). \tag{3.38}$$

Note that the definition of Σ_i^n in (3.31) leads to

$$\begin{pmatrix} B_i^n\Sigma_i^n \\ B_i^n\bar{A}\Sigma_i^n \\ \vdots \\ B_i^n\bar{A}^{q-1}\Sigma_i^n \\ B_i^n\bar{A}^q\Sigma_i^n \end{pmatrix} = \begin{pmatrix} H^n \\ H^n\bar{A} \\ \vdots \\ H^n\bar{A}^{q-1} \\ H^n\bar{A}^q \end{pmatrix} \tag{3.39}$$

which implies

$$B_i^n\Sigma_i^n = H^n \tag{3.40}$$

and

$$\underbrace{\begin{pmatrix} B_i^n \\ B_i^n\bar{A} \\ \vdots \\ B_i^n\bar{A}^{q-1} \end{pmatrix}}_{\mathscr{Q}_{B_i^n}} \bar{A}\Sigma_i^n = \underbrace{\begin{pmatrix} H^n \\ H^n\bar{A} \\ \vdots \\ H^n\bar{A}^{q-1} \end{pmatrix}}_{\mathscr{Q}_{H^n}} A \Leftrightarrow \bar{A}\Sigma_i^n = \Sigma_i^n\bar{A}. \tag{3.41}$$

It then follows from (3.20), (3.28) and (3.32) that

$$B_i\Sigma_i = H \tag{3.42}$$

and

$$A\Sigma_i = \Sigma_i A. \tag{3.43}$$

Using (3.21), (3.22), (3.33), (3.42) and (3.43), we obtain

$$\dot{\bar{\varpi}}_i = A\bar{\varpi} \tag{3.44}$$

and

$$B_i\bar{\varpi}_i = H\eta = v(t), \tag{3.45}$$

which ensure that $\bar{\varpi}_i$ is a solution to (3.37) such that $\hat{v}_i(t)$ in (3.38) is equal to $v(t)$ in (3.22).

We now prove the uniqueness of $\bar{\varpi}_i$ by contradiction. Suppose that there exists another solution $\hat{\varpi}_i(t)$ to (3.37) so that $\hat{v}_i(t) = v(t)$. Then let $\varpi_{e_i} = \hat{\varpi}_i - \bar{\varpi}_i$ and note that

$$\dot{\varpi}_{e_i} = A\varpi_{e_i} \tag{3.46}$$

and

$$B_i\varpi_{e_i} = 0. \tag{3.47}$$

It follows from the observability of $(\bar{A}, B_i^n)$ that (A, B_i) is observable, which implies from (3.47) that $\varpi_{e_i} \equiv 0$. Thus, $\bar{\varpi}_i$ must be unique. □

Proof (Theorem 3.1). We consider the storage functions $V_f(z)$ and $V_b(\xi)$ in (2.35) and the following storage function

$$V_p(\tilde{\varpi}) = \frac{1}{2}\tilde{\varpi}^T\tilde{\varpi}. \tag{3.48}$$

To compute the time derivative of $V_f(z)$, we define

$$\tilde{v}_i(t) = \hat{v}_i(t) - v(t), \quad i = 2, \cdots, N, \tag{3.49}$$

and note from (3.22), (3.30), (3.33) and (3.42) that

$$\tilde{v}_i(t) = B_i(\varpi_i(t) - \Sigma_i\eta(t)) = B_i\tilde{\varpi}_i. \tag{3.50}$$

For consistency, we let $\tilde{v}_1(t) \equiv 0_p$. We obtain from (2.6), (3.23) and (3.24)

$$\begin{aligned} \dot{z} &= (D^T \otimes I_p)\dot{x} && (3.51)\\ &= (D^T \otimes I_p)\{1_N \otimes v(t) + y + \tilde{v}\} && (3.52) \end{aligned}$$

where

$$\tilde{v} = [\tilde{v}_1^T, \tilde{v}_2^T, \cdots, \tilde{v}_N^T]^T. \tag{3.53}$$

Noting the fact

$$(D^T \otimes I_p)(1_N \otimes v(t)) = 0, \tag{3.54}$$

which follows from the third item in Property 1.5, we rewrite (3.52) as

$$\dot{z} = (D^T \otimes I_p)(y + \tilde{v}). \tag{3.55}$$

We use (3.55) and compute $\dot{V}_f$ as

$$\begin{aligned} \dot{V}_f &= \psi^T(D^T \otimes I_p)(y + \tilde{v})\\ &= \{(D \otimes I_p)\psi\}^T(y + \tilde{v})\\ &= -u^T(y + \tilde{v}). \end{aligned} \tag{3.56}$$

The time derivative of $V_b(\xi)$ is the same as (2.40).

Having computed $\dot{V}_f(z)$ and $\dot{V}_b(\xi)$, we now proceed to computing $\dot{V}_p(\tilde{\varpi})$. From (3.34), (3.29), (3.30), (3.33), (3.21), (3.22) and (3.43), we obtain

$$\begin{aligned} \dot{\tilde{\varpi}}_i &= A\varpi_i + B_i^T u_i - \Sigma_i A\eta && (3.57)\\ &= A(\varpi_i - \Sigma_i\eta) + B_i^T u_i && (3.58)\\ &= A\tilde{\varpi}_i + B_i^T u_i && (3.59) \end{aligned}$$

which, together with the passivity result in Lemma 3.1 and (3.50), results in

$$\dot{V}_p = \sum_{i=2}^{N} u_i^T B_i \tilde{\varpi}_i \tag{3.60}$$

$$= u^T \tilde{v}. \tag{3.61}$$

Thus, the time derivative of $V = V_f + V_b + V_p$ yields

$$\dot{V} = -\sum_{i \in \mathscr{I}} W_i(\xi_i) - \sum_{i \notin \mathscr{I}} u_i^T y_i \leq 0 \tag{3.62}$$

which proves the stability of (3.36). Because the closed-loop system (2.11), (3.55) and (3.59) is time invariant, we apply the Invariance Principle and analyze the largest invariant set $\mathscr{M}$ where $\dot{V} = 0$. It then follows that $u_i = 0$, $\forall i \notin \mathscr{I}$ and that $\xi_i = 0$, $\forall i \in \mathscr{I}$. We further note from (2.12) that $\xi_i = 0$, $i \in \mathscr{I}$ implies $u_i = 0$, which, together with $u_i = 0$, $\forall i \notin \mathscr{I}$, proves the convergence of the trajectories $(z(t), \xi(t), \tilde{\varpi}(t))$ to $\mathscr{E} \times \mathbb{R}^{qp(N-1)}$, where $\mathscr{E}$ is as in (2.33). In particular, when Property 2.1 holds, $u = 0$ means $z_k \in \mathscr{A}_k$ and thus all trajectories $(z(t), \xi(t), \tilde{\varpi}(t))$ starting in $\mathscr{G} \times \mathbb{R}^{pq(N-1)}$ converge to the set $\mathscr{A} \times \mathbb{R}^{pq(N-1)}$. □

The proof of Theorem 3.1 shows that the trajectories $(z(t), \xi(t), \tilde{\varpi}(t))$ converge to an invariant set $\mathscr{M}$ where $\xi = 0$ and $u = 0$, which means that objective A2 is achieved at least locally and that $|\dot{x}_1 - v(t)| \to 0$ and $|\dot{x}_i - \hat{v}_i(t)| \to 0$, $i = 2, \cdots, N$. If, in addition, $\hat{v}_i$ converges to $v(t)$, the agents all reach the reference velocity, thereby guaranteeing objective A1. The next corollary characterizes two possible situations where objective A1 is ensured.

Corollary 3.1. *Suppose that the assumptions in Theorem 3.1 hold. If, in addition, one of the following conditions holds:*

1. *$\bar{A}$ in (3.15) is a zero matrix;*
2. *Property 2.1 holds and $\mathscr{A}_k$ is a point, $\forall k = 1, \cdots, \ell$,*

then all trajectories $(z(t), \xi(t), \tilde{\varpi}(t))$ converge to the equilibria set $\mathscr{E}^*$ in (3.35). In particular, $\tilde{\omega} \to 0$ and $|\hat{v}_i(t) - v(t)| \to 0$, $i = 2, \cdots, N$. Moreover, if condition 2 holds, the set $\mathscr{A}^*$ in (3.36) is uniformly asymptotically stable with the region of attraction $\mathscr{G} \times \mathbb{R}^{pq(N-1)}$. □

If condition 1 is satisfied in Lemma 3.1, the reference velocity in (3.21)-(3.22) is constant and $\hat{v}_i(t)$ then acts as an integral control that eliminates the constant disturbance due to the different reference velocity information. An example of condition 2 is the agreement problem, where $\mathscr{A}_k = 0$ and Property 2.1 is satisfied due to (2.46). Note that condition 2 restricts the desired target sets $\mathscr{A}_k$ but not $v(t)$ while condition 1 restricts $v(t)$ instead of $\mathscr{A}_k$.

Proof (Corollary 3.1). Noting the results in Theorem 3.1, we only need to prove $\tilde{\omega} \to 0$ and $|\hat{v}_i(t) - v(t)| \to 0$, $i = 2, \cdots, N$ for both conditions. For condition 1, $v(t)$ in (3.22) is constant since A in (3.21) is a zero matrix. Because $u = 0$ on $\mathscr{M}$, we conclude from (3.29) and (3.30) that $\hat{v}_i(t)$ is constant on $\mathscr{M}$. For dynamic block $\mathscr{H}_i$, we obtain from $\xi_i = 0$, $u_i = 0$, (2.11) and (2.13) that $y_i = 0$. Likewise, the static block satisfies $y_i = h_i(u_i)$, which implies that y_i vanishes on $\mathscr{M}$. Using (3.23) and (3.24), we have $\dot{x}_1 = v$ and $\dot{x}_i = \hat{v}_i$, where v and $\hat{v}_i$ are constant. Since $\dot{z} = (D^T \otimes I_p)\dot{x}$, $\dot{z}$ is also constant and thus the only way that $z(t)$ can stay in the bounded invariant set $\mathscr{M}$ is when $\dot{z} = 0$. We then note from (3.55) that $(D^T \otimes I_p)\tilde{v} = 0$. Recalling that $\tilde{v}_1 \equiv 0$, we conclude $\tilde{v} = 0$ and $|\hat{v}_i - v(t)| \to 0$, $i = 2, \cdots, N$. The convergence $\tilde{\omega} \to 0$ follows from the uniqueness property in Lemma 3.2.

For condition 2, note that since Property 2.1 is satisfied, z is constant on $\mathscr{M}$. Thus, $\dot{z} = 0$. Following an analysis similar to that for condition 1, we obtain $|\hat{v}_i - v(t)| \to 0$, $i = 2, \cdots, N$ and $\tilde{\omega} \to 0$. □

3.4 Design Examples for Distance-based Formation Control

In this section, we illustrate two examples of the basic adaptive design: The first one assumes a constant $v(t)$, and the second one shows that for a time-varying $v(t)$, $\hat{v}_i$ may fail to converge to $v(t)$, which motivates us to redesign the feedback (2.23) in Section 3.5.

We consider the formation control of three agents, where agent 1 is the leader, which means that $v(t)$ is available only to agent 1. In this example, we assume that the agent model is given by (2.16) with $m_i = 1$, $i = 1, 2, 3$. The desired formation is an equilateral triangle with side length 1.

The dynamics of agent 1 remain the same as in (2.19) and (2.20). For the agents $i = 2, 3$, τ_i in (2.17) must be replaced with

$$\tau_i = -k_i(\dot{x}_i(t) - \hat{v}_i(t)) + m_i \dot{\hat{v}}_i(t) + u_i, \tag{3.63}$$

which, together with a change of variables $\xi_i = \dot{x}_i - \hat{v}_i$, brings the dynamics of the agents, $i = 2, 3$ to the form

$$\dot{x}_i = \xi_i + \hat{v}_i \tag{3.64}$$

$$m_i \dot{\xi}_i = -k_i \xi_i + u_i \tag{3.65}$$

where u_i is as in (2.23), $\psi_k(z_k)$ is as in (2.87) and $\sigma_k(\cdot)$ is taken as the natural logarithm. The signal $\hat{v}_i$ is available for implementation in (3.64) once the update law in (3.29)-(3.30) is setup. The graph G is the same as in Section 2.6.2.

3.4.1 Constant Reference Velocity

We select agent 1 to possess a constant reference velocity $v = [0.2 \;\; 0.2]^T$. It follows that A in (3.21) is $\mathbf{0}_2$. We choose $\eta(0) = [0.2 \;\; 0.2]^T$ and $H = I_2$ in (3.21)-(3.22). Agents 2 and 3 implement (3.63) and update $\hat{v}_i$ according to (3.29)-(3.30) with $A = \mathbf{0}_2$ and $B_i = H$. In Fig. 3.2, three agents first start with the adaptation turned off. Since agents 2 and 3 possess incorrect information about $v(t)$ and since there is no adaptation, the relative distances $|z_k|$ do not converge to the prescribed sets $\mathscr{A}_k$. However, we note that $|z_k|$'s remain bounded because the interaction forces $\psi_k(z_k)$ act as "spring forces" between neighboring agents and prevent them from diverging. At $t = 10$ sec, we turn on the adaptation for agents 2 and 3, which results in convergence to the desired distances $|z_k| = 1$ asymptotically. At the same time, Fig. 3.3 illustrates that the estimated velocities $\hat{v}_i^1$ and $\hat{v}_i^2$ converge to the corresponding reference velocities v_1^1 and v_1^2, as guaranteed by Corollary 3.1.

3.4.2 Motivating Example for the Augmented Design

In this example, $v(t)$ in the x_1-dynamics (2.19) is chosen to be time-varying and satisfies (3.21)-(3.22) with

$$A = I_2 \otimes \begin{pmatrix} 0 & 1 \\ -1 & 0 \end{pmatrix} \tag{3.66}$$

and

$$H = \frac{1}{2}\begin{pmatrix} 1 & 1 & 0 & 0 \\ 0 & 0 & -1 & 1 \end{pmatrix}. \tag{3.67}$$

We pick the initial condition of η as $\eta(0) = [\frac{\sqrt{3}}{3} \ -\frac{\sqrt{3}}{3} \ -\frac{\sqrt{3}}{3} \ \frac{\sqrt{3}}{3}]^T$. It is easy to verify that $v(t) = [-\frac{\sqrt{3}}{3}\sin(t) \ \frac{\sqrt{3}}{3}\cos(t)]^T$, which implies that x_1 will rotate with a radius of $\frac{\sqrt{3}}{3}$. The other agents estimate $v(t)$ by (3.29)-(3.30) with A in (3.66) and $B_i = H$ in (3.67), $i = 2,3$.

To show the lack of estimate convergence, we suppose that initially the three agents form an equilateral triangle where $x_1(0) = [\frac{\sqrt{3}}{3} \ 0]^T$, $x_2(0) = [-\frac{\sqrt{3}}{6} \ \frac{1}{2}]^T$ and $x_3(0) = [-\frac{\sqrt{3}}{6} \ -\frac{1}{2}]^T$, thus satisfying the desired formation with $|z_k| = 1$, $k = 1,2,3$. Fig. 3.4(a) shows that in the nonadaptive design where the reference velocity $v(t)$ is available to each agent, the group exhibits a translational motion with x_1 spinning around the origin. The adaptive case in Fig. 3.4(b) where the initial conditions are set to $\xi_1(0) = \xi_2(0) = \xi_3(0) = 0$, $\varpi_2(0) = [-\frac{1}{2} - \frac{\sqrt{3}}{6} \ \frac{\sqrt{3}}{6} - \frac{1}{2} \ -\frac{1}{2} + \frac{\sqrt{3}}{6} \ \frac{1}{2} - \frac{\sqrt{3}}{6}]^T$ and $\varpi_3(0) = [\frac{1}{2} - \frac{\sqrt{3}}{6} \ \frac{\sqrt{3}}{6} + \frac{1}{2} \ \frac{\sqrt{3}}{6} - \frac{1}{2} \ -\frac{\sqrt{3}}{6} - \frac{1}{2}]^T$, shows that the agents $i = 2,3$, exhibit a rotational motion about the leader, which means that the $v(t)$ information is not fully recovered for agents 2 and 3. However, the agents still maintain the group formation as expected from Theorem 3.1.

To recover velocity tracking of a time-varying $v(t)$, one choice is to modify the sets $\mathscr{A}_k$ to the formulation of the position-based formation control in (2.51). Since position-based formation control can be transformed to the agreement problem, condition 2 in Corollary 3.1 is satisfied and thus $|\hat{v}_i(t) - v(t)| \to 0$ is guaranteed. Fig. 3.5 shows that the basic adaptive design achieves the same group behavior as in the nonadaptive design Fig. 3.4(a), by taking the desired target sets to be the same as (2.81) with the choice of the interaction forces

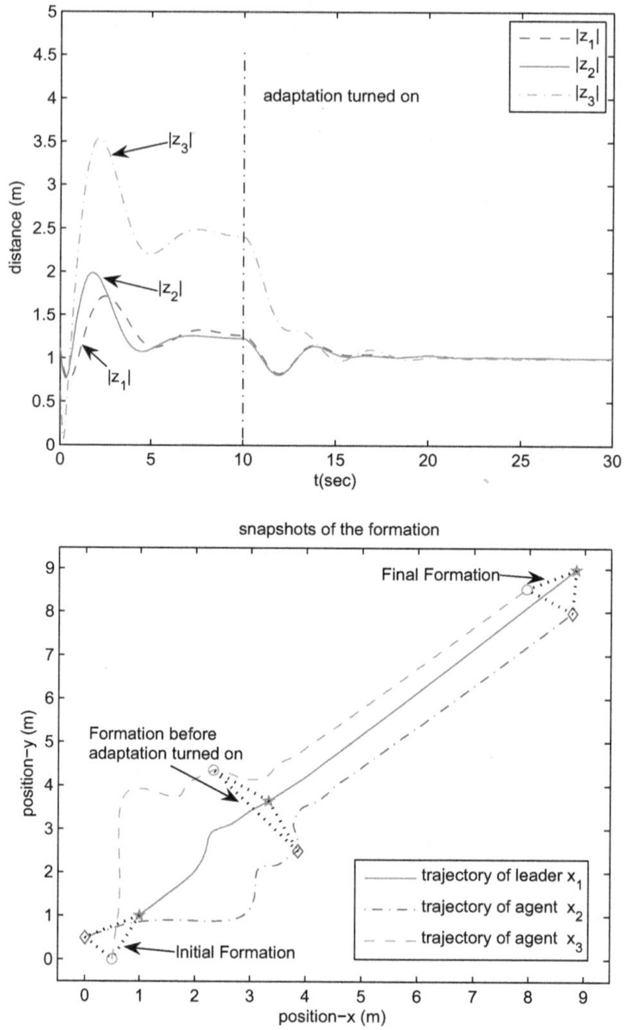

Fig. 3.2 Snapshots of the formation in the adaptive design with constant $v(t)$: In the first 10 seconds, the adaptation is off and the desired formation is not achieved. The adaptation is turned on at $t = 10$ sec, after which point the trajectories converge to the desired formation. [11]. Reprinted with Permission. Copyright Elsevier 2008.

$$\psi_k(z_k) = z_k - z_k^d.$$

In the next section, we show that full recovery of the $v(t)$ information is ensured by augmenting u_i with relative velocity information. This augmented design does not need to modify the desired target set $\mathscr{A}_k$.

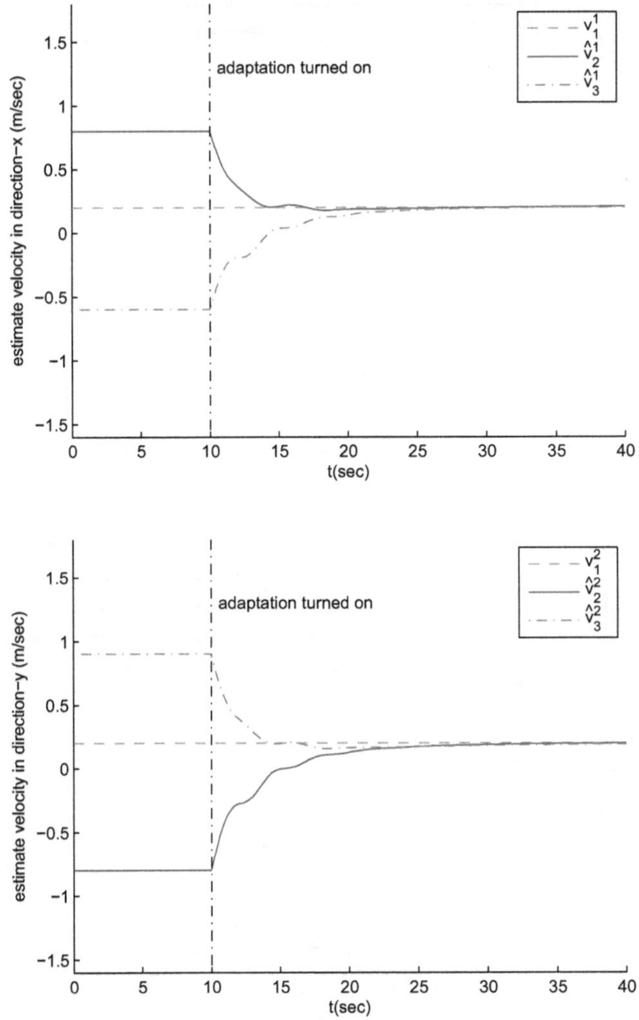

Fig. 3.3 Velocity convergence with the adaptive scheme in Fig. 3.2. [11]. Reprinted with Permission. Copyright Elsevier 2008.

3.5 The *Augmented* Design

We now develop an augmented design that guarantees tracking of the reference velocity $v(t)$. In the basic update law (3.29)-(3.30), $\hat{v}_i$ stops updating when u_i in (2.23) reaches zero. As shown in Section 3.4.2, however, $u_i = 0$ does not mean that all the agents possess the same velocity $v(t)$. Note that, in contrast, $\dot{z} \to 0$ would

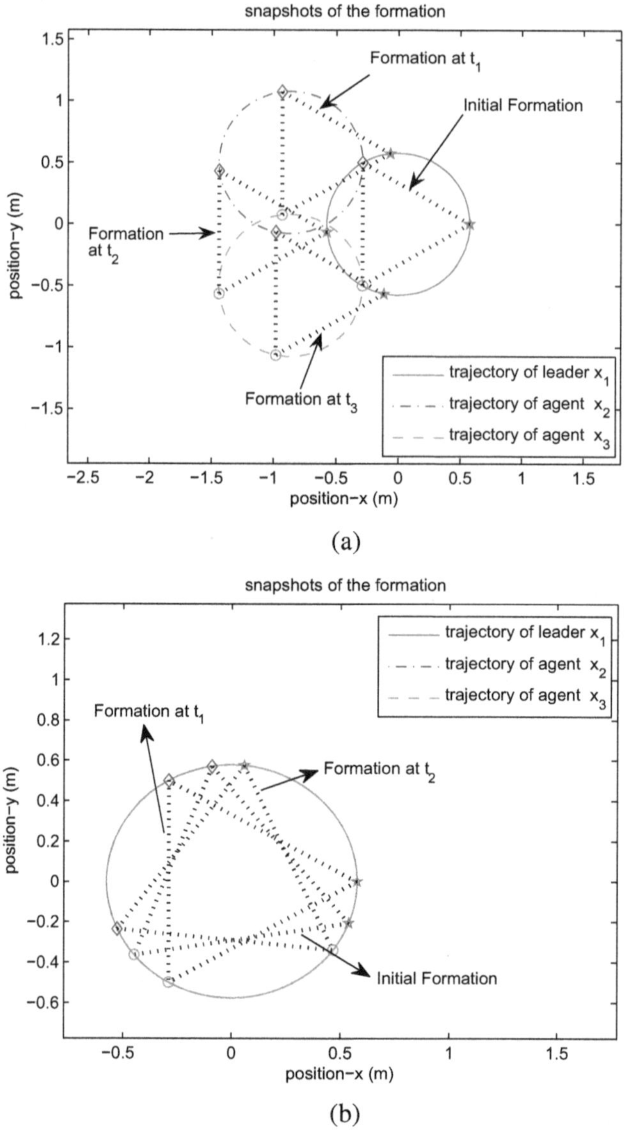

Fig. 3.4 Two group behaviors: (a) Nonadaptive design with time-varying $v(t)$: The group exhibits a translational motion with x_1 spinning around the origin. (b) Adaptive design with time-varying $v(t)$: The agents x_2 and x_3 exhibit a rotational motion about the leader x_1.

imply that all agents converge to the same velocity. Thus, in the augmented adaptive design we employ $\dot{z}$ to guarantee tracking of the reference velocity.

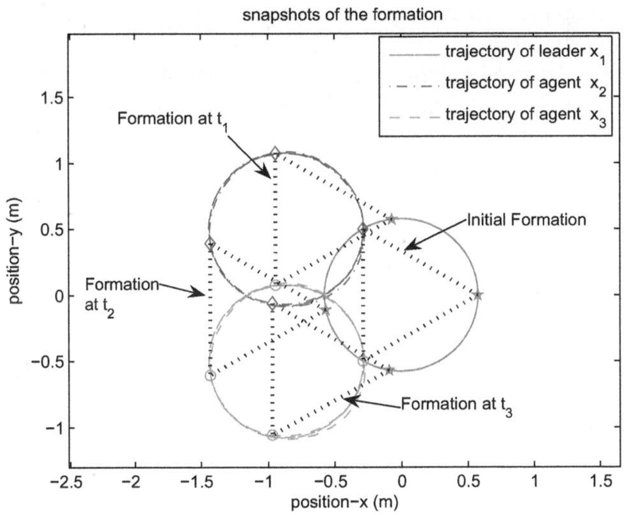

Fig. 3.5 Adaptive design with modified desired sets. The group exhibits the same translational motion as in the nonadaptive design Fig. 3.4(a).

To present the augmented design, we introduce a static *directed* graph G^v representing the information topology for the relative velocity: If the ith agent has access to the relative velocity information $\dot{x}_i - \dot{x}_j$, then the nodes i and j in the graph G^v are connected by a directional link from j to i and agent j is a neighbor of agent i. We denote by $\mathscr{N}_i^v$ the set of neighbors of agent i in G^v. We assume that G^v is balanced and strongly connected.

Instead of the external feedback in (2.23), we now propose the augmented design

$$u_i = -\sum_{k=1}^{\ell} d_{ik}\psi_k(z_k) - \sum_{j\in\mathscr{N}_i^v} (\dot{x}_i - \dot{x}_j) \tag{3.68}$$

with the update law (3.29)-(3.30). This augmented design recovers the stability result of Theorem 2.1 and ensures tracking of the reference velocity.

Theorem 3.2. *Consider the coordination laws in (3.23) and (3.24) which are designed the same as in Theorem 3.1 except that u_i is now defined in (3.68). Assume that G^v is static, strongly connected and balanced. Then, the desired equilibria set $\mathscr{A}^*$ in (3.36) is stable. All trajectories $(z(t), \xi(t), \tilde{\varpi}(t))$ starting in $\mathscr{G} \times \mathbb{R}^{pq(N-1)}$ are bounded and converge to the equilibria set $\mathscr{E}^*$ in (3.35). If, in addition, Property 2.1 holds, all trajectories $(z(t), \xi(t), \tilde{\varpi}(t))$ starting in $\mathscr{G} \times \mathbb{R}^{pq(N-1)}$ converge to $\mathscr{A}^*$ in (3.36).* □

Proof. Using (1.11), (3.24), and (3.49), we rewrite (3.68) in the compact form

$$u = -(D\otimes I_p)\psi - (L^v \otimes I_p)(y + \tilde{v}) \tag{3.69}$$

where L^v is the graph Laplacian matrix of the velocity graph G^v and we have used $L^v 1_N = 0$ in Property 1.1.

We take the same storage functions as in (2.35) and (3.48). With the new feedback (3.68), we note that $\dot{V}_b$ and $\dot{V}_p$ remain the same as in (2.40) and (3.61) while $\dot{V}_f$ is now given by

$$\dot{V}_f = (-u - (L^v \otimes I_p)(y + \tilde{v}))^T (y + \tilde{v}). \tag{3.70}$$

Thus, the time derivative of $V = V_p + V_f + V_b$ is

$$\dot{V} = -\sum_{i \in \mathscr{I}} W_i(\xi_i) - \sum_{i \notin \mathscr{I}} u_i^T y_i - (y + \tilde{v})^T (L^v \otimes I_p)(y + \tilde{v}). \tag{3.71}$$

When G^v is balanced and strongly connected, $(y + \tilde{v})^T (L^v \otimes I_p)(y + \tilde{v})$ is nonnegative (see Property 1.4). Thus, the derivative in (3.71) is negative semidefinite, which implies global stability of $\mathscr{A}^*$ in (3.36) and boundedness of all the signals $(z(t), \xi(t), \tilde{\varpi}(t))$.

We next apply the Invariance Principle and investigate the largest invariant set $\mathscr{M}'$ where $\dot{V} = 0$. It follows that $\xi_i = 0$, $\forall i \in \mathscr{I}$, $u_i = 0$, $\forall i \notin \mathscr{I}$ and $(y + \tilde{v})^T (L^v \otimes I_p)(y + \tilde{v}) = 0$. Applying the same analysis as in the proof of Theorem 3.1, we conclude $u = 0$ on $\mathscr{M}'$. For dynamic block $\mathscr{H}_i$, we obtain from $\xi_i = 0$, $u_i = 0$, (2.11) and (2.13) that $y_i = 0$. Likewise, the static block satisfies $y_i = h_i(u_i)$, which implies that y_i vanishes on $\mathscr{M}'$. Therefore, $y_i = 0$ on $\mathscr{M}'$, $\forall i$. Recall from Property 1.4 that $(y + \tilde{v})^T (L^v \otimes I_p)(y + \tilde{v})$ is zero only when $(y + \tilde{v})^T (L^v_{\text{sym}})(y + \tilde{v})$ is zero. Since the graph G^v is strongly connected, the graph corresponding to L^v_{sym} is connected. Therefore, $(y + \tilde{v})^T (L^v \otimes I_p)(y + \tilde{v}) = 0$ implies $y + \tilde{v} = 1_N \otimes c$, where $c \in \mathbb{R}^p$. We then conclude from $\tilde{v}_1 \equiv 0$ and $y_1 = 0$ that $|y + \tilde{v}| = 0$. Since $y = 0$, it follows that $\tilde{v} = 0$, which implies from (3.49) that tracking of $v(t)$ is achieved. Because of the uniqueness property in Lemma 3.2, $\tilde{\varpi} \to 0$ follows. □

3.5.1 Motivating Example Revisited

We now include the relative velocity information in the external feedback u_i in the adaptive design in Section 3.4.2. We assume that G^v is a directed cyclic graph, where agent i is the only neighbor of agent $i+1$ for $i = 1, 2$ and agent 3 is the only neighbor of agent 1. We modify u_i in Section 3.4.2 by adding the relative velocity for each agent with respect to its neighbor. Fig. 3.6 shows the snapshots of the formation. The group now exhibits a translational motion with x_1 circling around the origin, which means that the nonadaptive results are fully recovered. In addition, the estimates of the reference velocity converge to their true values as shown in Fig. 3.7.

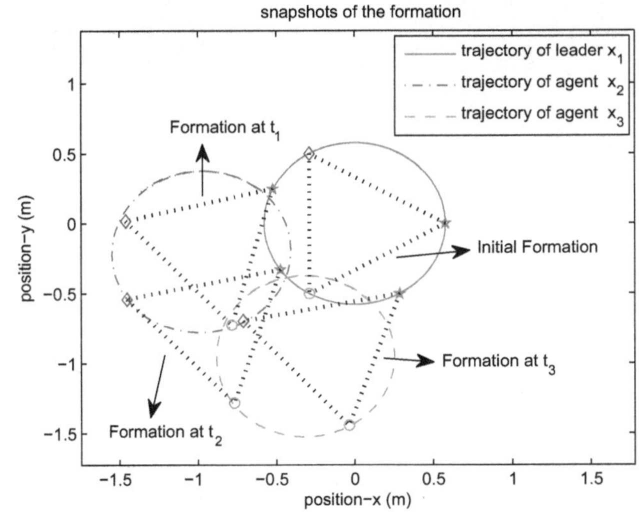

Fig. 3.6 The augmented adaptive design (3.68) recovers the velocity tracking of the nonadaptive design as well as ensuring the desired formation.

3.6 When There Is No Leader

In the case where there is no leader in the group and no prescribed reference velocity $v(t)$ is assigned to the group, we let each agent implement

$$\dot{x}_i = y_i + \hat{v}_i(t), \quad i = 1, \cdots, N, \tag{3.72}$$

where $\hat{v}_i(t)$ is updated by (3.29)-(3.30). When u_i in (3.29) is given by (2.23), Corollary 3.2 below proves the same convergence results as in Theorem 3.1. In particular, if condition 1 or 2 in Corollary 3.1 holds, the design (3.72) guarantees that $\dot{x}_i$ converges to a common velocity $\bar{v}(t)$, $\forall i = 1, \cdots, N$. When $B_i = B$ in (3.29) for some B matrix, $\forall i$, this common velocity $\bar{v}(t)$ is characterized by

$$\dot{\chi} = A\chi \tag{3.73}$$

$$\bar{v}(t) = B\chi \tag{3.74}$$

where $\chi(0) = \frac{1}{N}\sum_{i=1}^{N} \varpi_i(0)$.

Corollary 3.2. *Consider the coordination law in (3.72), where $\hat{v}_i(t)$ is given by (3.29)-(3.30), u_i is defined in (2.23) in which ψ_k, $k = 1, \cdots, \ell$, are designed as in (2.27)-(2.31), and y_i is the output of $\mathcal{H}_i$ with the input u_i, $i = 1, \cdots, N$, in which the passive block $\mathcal{H}_i$ is designed as in (2.11)-(2.15). Then, all trajectories $(z(t), \xi(t), \varpi(t))$ starting in $\mathcal{G} \times \mathbb{R}^{pqN}$ are bounded and converge to the set $\mathcal{E} \times \mathbb{R}^{pqN}$, where $\varpi = [\varpi_1^T, \cdots, \varpi_N^T]^T$ and $\mathcal{E}$ and $\mathcal{G}$ are as in (2.33) and (2.37).*

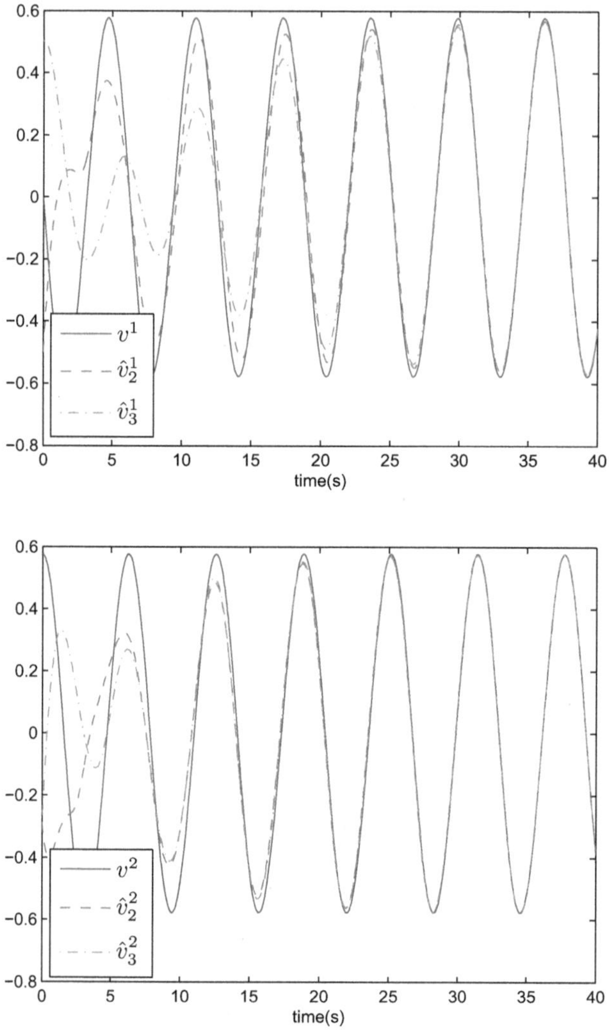

Fig. 3.7 With the augmented adaptive design (3.68), the agents' estimates converge to the true reference velocity $v(t)$.

When Property 2.1 holds, all trajectories $(z(t), \xi(t), \varpi(t))$ *starting in* $\mathscr{G} \times \mathbb{R}^{pqN}$ *converge to the set* $\mathscr{A} \times \mathbb{R}^{pqN}$*, where* $\mathscr{A}$ *is as in (2.36).*

When condition 1 *or* 2 *in Corollary 3.1 holds, there exists a bounded* $\bar{v}(t) \in \mathbb{R}^p$ *such that* $|\hat{v}_i - \bar{v}(t)| \to 0$*,* $\forall i$*. In addition, if* $B_i = B$ *in (3.29),* $\forall i$*, then* $\bar{v}(t)$ *is as in (3.73)-(3.74) and* $|\varpi_i(t) - \chi(t)| \to 0$*.* □

Likewise, when u_i is augmented with relative velocity information as in (3.68), $\dot{x}_i$'s reach agreement without requiring either condition in Corollary 3.1:

Corollary 3.3. *Consider the coordination law in (3.72), which is designed the same as in Corollary 3.2 except that u_i is now defined in (3.68). Suppose that G^v is constant, strongly connected and balanced. Then, all trajectories $(z(t), \xi(t), \varpi(t))$ starting in $\mathcal{G} \times \mathbb{R}^{pqN}$ are bounded, where $\mathcal{G}$ is in (2.37). The signals $(z(t), \xi(t))$ converge to the set $\mathcal{E}$, where $\mathcal{E}$ is as in (2.33). When Property 2.1 holds, $(z(t), \xi(t))$ converge to the set $\mathcal{A}$, where $\mathcal{A}$ is as in (2.36). Moreover, $|\hat{v}_i - \bar{v}(t)| \to 0$ for some bounded $\bar{v}(t) \in \mathbb{R}^p$ and for all i. If $B_i = B$ in (3.29), $\forall i$, then $\bar{v}(t)$ is as in (3.73)-(3.74) and $|\varpi_i(t) - \chi(t)| \to 0$.* □

The proofs for Corollary 3.2 and 3.3 are given in Appendix A.1 and A.2, respectively.

Chapter 4
Adaptive Design for Reference Velocity Recovery: Parameterization Approach

4.1 Introduction

The designs in Sections 3.3 and 3.5 restrict the reference velocity $v(t)$ to be constant or periodically time-varying. In this section, we present adaptive designs that are applicable to any time-varying, uniformly bounded and C^1 reference velocity $v(t)$ that can be parameterized as

$$v(t) = \sum_{j=1}^{r} \phi^j(t)\theta^j = (\Phi(t)^T \otimes I_p)\theta \tag{4.1}$$

where $\phi^j(t) \in \mathbb{R}$, $j = 1, \cdots, r$ are basis functions available to each agent, $\theta^j \in \mathbb{R}^p$ are column vectors available only to the leader,

$$\Phi(t) = [\phi^1(t), \cdots, \phi^r(t)]^T \tag{4.2}$$

and

$$\theta = [(\theta^1)^T, \cdots, (\theta^r)^T]^T. \tag{4.3}$$

We let agent i, $i = 2, \cdots, N$, estimate the unknown θ^j by $\hat{\theta}_i^j$, and construct $\hat{v}_i(t)$ from

$$\hat{v}_i(t) = \sum_{j=1}^{r} \phi^j(t)\hat{\theta}_i^j = (\Phi(t)^T \otimes I_p)\hat{\theta}_i \quad i = 2, \cdots, N, \tag{4.4}$$

where

$$\hat{\theta}_i = [(\hat{\theta}_i^1)^T, \cdots, (\hat{\theta}_i^r)^T]^T. \tag{4.5}$$

In the following sections, we first develop a basic adaptive design with which agent i updates its estimate $\hat{\theta}_i$, $i = 2, \cdots, N$. Like the design in Section 3.3, this basic adaptive design recovers objective A2 as well as guaranteeing objective A1 in some special cases, such as, the agreement problem. To ensure objective A1 when the basic adaptive design fails, we then modify the basic adaptive design in a similar fashion to Section 3.5 and obtain the augmented adaptive design. We next apply the adaptive design result to an extremum seeking example. In this example, a group leader autonomously determines the Newton direction towards the extremum by sampling a field distribution and parameterizes the group reference velocity according to the Newton direction. The other agents then estimate this reference velocity using the basic adaptive design and reconstruct the desired formation during extremum seeking. Before proceeding to these results, we first compare the parameterization approach and the internal model approach.

In the parameterization approach, the availability of the basis functions to each agent is similar to Assumption 1 in the internal model approach, where the $\bar{A}$ matrix is available to each agent. The use of the basis functions $\phi^j(t)$ removes the restriction in the internal model approach that $v(t)$ be constant or periodic. The basis functions may then be used to shape the transient of the reference velocity profile. However, since $\phi^j(t)$'s are time-dependent, the agents need to have synchronized clocks to implement this parameterization approach. The next example compares the number of the internal states used for estimating periodic reference velocity in these two approaches.

Example 4.1. We consider a scalar reference velocity $v(t)$, parameterized by

$$v(t) = \sum_{i=1}^{r_1} (a_i \sin(w_i t) + b_i \cos(w_i t)). \tag{4.6}$$

Note that $v(t)$ in (4.6) is already parameterized by the basis functions $\sin(w_i t)$ and $\cos(w_i t)$, $i = 1, \cdots, r_1$. Therefore, the total number of unknown parameters that parameterizes this $v(t)$ is $2r_1$, which means that for the parameterization approach, each agent (except the leader) needs to update $2r_1$ internal states to estimate these unknown parameters.

In the internal model approach, we choose $\bar{A}$ in (3.15) as

$$\bar{A} = \text{diag}\left\{ \begin{pmatrix} 0 & -w_1 \\ w_1 & 0 \end{pmatrix}, \cdots, \begin{pmatrix} 0 & -w_{r_1} \\ w_{r_1} & 0 \end{pmatrix} \right\} \tag{4.7}$$

which implies that the dimension of ϖ_i in (3.29) is $2r_1$. This means that each agent (except the leader) also maintains $2r_1$ internal states to estimate the reference velocity. Thus, in estimating generic periodic reference velocities, both the parameterization approach and the internal model approach use the same number of internal states.

In some special cases, however, the parameterization approach may require less internal states. For example, if in (4.6) $b_i = 0$, $i = 1, \cdots, r_1$, then the parameterization approach only requires r_1 internal states which estimate all a_i's while the internal model approach still requires $\bar{A}$ to be the same as (4.7) and the dimension of ϖ_i to be $2r_1$. □

4.2 The Basic Design

We choose the update law for the parameter $\hat{\theta}_i$ in (4.4) as

$$\dot{\hat{\theta}}_i = \Lambda_i(\Phi(t)\otimes I_p)u_i \tag{4.8}$$

in which $\Lambda_i = \Lambda_i^T > 0$ and u_i is as in (2.23). As proven in Theorem 4.1 below, the basic adaptive design (3.23), (3.24), and (4.8) guarantees convergence to the desired target sets (objective A2). Whether objective A1 is achieved or not depends on the convergence of $\hat{\theta}_i$ to θ, which will be studied in Section 4.3. When $\hat{\theta}_i$ converges to θ, $v(t)$ is recovered with the adaptive design and, thus, object A1 is also achieved.

Theorem 4.1. *Consider the coordination laws in (3.23), (3.24), (4.4) and (4.8) where $v(t)$ is uniformly bounded and piecewise continuous, parameterized as (4.1) in which $\phi^j(t)$, $j = 1,\cdots,r$ are uniformly bounded, and $\mathscr{H}_i$, $i = 1,\cdots,N$, and ψ_k, $k = 1,\cdots,\ell$ are designed as in (2.11)-(2.15) and (2.27)-(2.31), respectively. Then, the set*

$$\mathscr{E}^* = \left\{(z,\xi,\hat{\theta})|\ \xi = 0,\ (D\otimes I_p)\psi(z) = 0 \text{ and } z\in\mathscr{R}(D^T\otimes I_p),\ \hat{\theta} = \theta^*\right\} \tag{4.9}$$

is stable, where $\hat{\theta} = [\hat{\theta}_2^T,\cdots,\hat{\theta}_N^T]^T$ and $\theta^* = 1_{N-1}\otimes\theta$. All trajectories $(z(t),\xi(t),\hat{\theta}(t))$ starting in $\mathscr{G}\times\mathbb{R}^{pr(N-1)}$ are bounded and converge to the set $\mathscr{E}\times\mathbb{R}^{pr(N-1)}$, where $\mathscr{E}$ and $\mathscr{G}$ are as in (2.33) and (2.37). Moreover, when Property 2.1 holds, all trajectories $(z(t),\xi(t),\hat{\theta}(t))$ starting in $\mathscr{G}\times\mathbb{R}^{pr(N-1)}$ converge to the set $\mathscr{A}\times\mathbb{R}^{pr(N-1)}$, where $\mathscr{A}$ is as in (2.36). □

To obtain the closed-loop structure of the basic adaptive design, we denote by $\tilde{\theta}_i$ the error variable

$$\tilde{\theta}_i = \hat{\theta}_i - \theta \quad i = 2,\cdots,N, \tag{4.10}$$

and note from (4.8) that

$$\dot{\tilde{\theta}}_i = \Lambda_i(\Phi(t)\otimes I_p)u_i. \tag{4.11}$$

Using (4.1) and (4.4), we get

$$\tilde{v}_i = \hat{v}_i - v(t) = (\Phi(t)^T\otimes I_p)\tilde{\theta}_i, \quad i = 2,\cdots,N. \tag{4.12}$$

We set $\tilde{\theta}_1 \equiv 0$ and $\tilde{v}_1 \equiv 0$, and define

$$\tilde{\theta} = [\tilde{\theta}_1^T,\tilde{\theta}_2^T,\cdots,\tilde{\theta}_N^T]^T \tag{4.13}$$

and

$$\tilde{v} = (I_N\otimes\Phi^T(t)\otimes I_p)\tilde{\theta} = [\tilde{v}_1^T,\cdots,\tilde{v}_N^T]^T. \tag{4.14}$$

The closed-loop structure of the basic adaptive design is then shown in Fig. 4.1.

We now give a passivity interpretation of the basic adaptive design. Because the single integrator is passive and because the feedback path from u to $\tilde{v}$ exhibits the same structure as Structure 1, we obtain the passivity from u to $\tilde{v}$. We then conclude

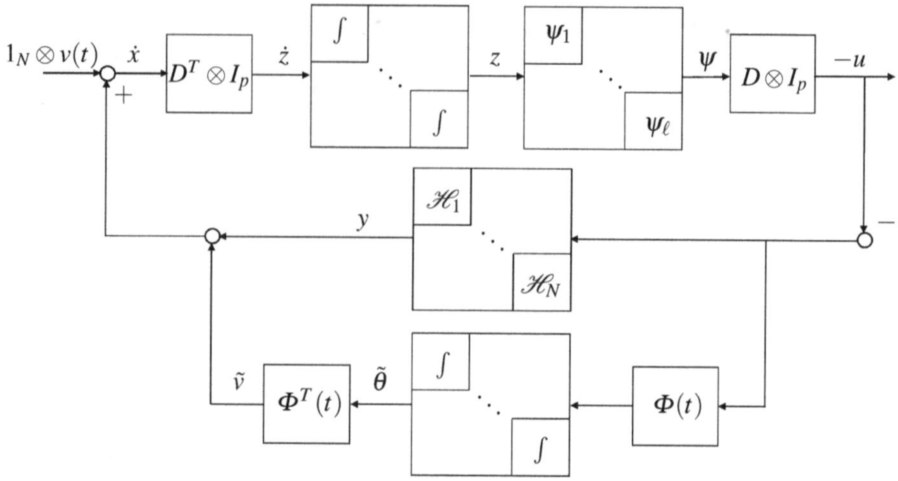

Fig. 4.1 The closed-loop structure of the basic adaptive design. The appearance of $\Phi(t)$ and its transpose before and after the integrator implies the passivity from u to $\tilde{v}$. The closed-loop stability follows from the interconnection of the passive feedforward path and the passive feedback paths.

from the passivity result *ii*) in Theorem 2.1 and Structure 2 that the feedback path is passive from u to $y+\tilde{v}$. As proven in Theorem 2.1, the feedforward path is also passive. Therefore, the closed-loop stability follows from Structure 3. The detailed proof is given below.

Proof. To prove the stability of the closed-loop system described by the adaptive design (2.11), (3.55) and (4.11), we exploit the passivity properties of the interconnected systems and consider $V_f(z)$ and $V_b(\xi)$ in (2.35) and

$$V_a(\tilde{\theta}) = \frac{1}{2}\sum_{i=2}^{N} \tilde{\theta}_i^T \Lambda_i^{-1} \tilde{\theta}_i, \tag{4.15}$$

which are the storage functions for the three paths in Fig. 4.1. In particular, the time derivatives of $V_f(z)$ and $V_b(\xi)$ are the same as (3.56) and (2.40).

Using (4.11), we obtain

$$\begin{aligned}
\dot{V}_a &= \sum_{i=2}^{N} \tilde{\theta}_i^T \Lambda_i^{-1} \dot{\tilde{\theta}}_i \\
&= \sum_{i=2}^{N} \tilde{\theta}_i^T \Lambda_i^{-1} \Lambda_i (\Phi(t) \otimes I_p) u_i \\
&= \sum_{i=2}^{N} \tilde{\theta}_i^T (\Phi(t) \otimes I_p) u_i \\
&= u^T \tilde{v}.
\end{aligned} \tag{4.16}$$

From (2.40), (3.56) and (4.16), the Lyapunov function

$$V(z,\xi,\tilde{\theta}) = V_f(z) + V_b(\xi) + V_a(\tilde{\theta}) \tag{4.17}$$

yields the negative semidefinite derivative

$$\dot{V} \leq -\sum_{i\in\mathscr{I}}^{N} W_i(\xi_i) - \sum_{i\notin\mathscr{I}} u_i^T y_i \leq 0 \tag{4.18}$$

which implies that all the trajectories $(z(t),\xi(t),\tilde{\theta}(t))$ are bounded. We further conclude from Theorem B.5 in Appendix B.3 that $\xi_i \to 0$, $\forall i \in \mathscr{I}$ and that $u_i \to 0$, $\forall i \notin \mathscr{I}$. We next show that $u_i \to 0$, $\forall i \in \mathscr{I}$. To this end we note that

$$\ddot{\xi}_i = \frac{\partial f_i}{\partial u_i}\dot{u}_i + \frac{\partial f_i}{\partial \xi}\dot{\xi}_i \tag{4.19}$$

is continuous and uniformly bounded because $\dot{u}$ and $\dot{\xi}$ are continuous functions of the bounded signals $(z(t),\xi(t),\tilde{\theta}(t),\Phi(t))$ and because $f_i(\cdot,\cdot)$ is C^1. Since $\xi_i \to 0$ and $\ddot{\xi}_i$ is continuous and bounded, it follows from Theorem B.4 in Appendix B.3 that $\dot{\xi}_i \to 0$, which, from (2.11) and (2.12), guarantees $u_i \to 0$.

Finally, we note that $u \to 0$ implies from (2.26) that $\psi(z)$ converges to the null space $\mathscr{N}(D\otimes I_p)$. This, in turn, implies that the trajectories $(z(t),\xi(t),\hat{\theta}(t))$ starting in $\mathscr{G}\times\mathbb{R}^{pr(N-1)}$ converge to the set $\mathscr{E}\times\mathbb{R}^{pr(N-1)}$, where $\mathscr{E}$ and $\mathscr{G}$ are as in (2.33) and (2.37). Moreover, when Property 2.1 holds, all trajectories converge to the set $\mathscr{A}\times\mathbb{R}^{pr(N-1)}$, where $\mathscr{A}$ is as in (2.36). □

4.3 Parameter Convergence

Parameter convergence is essential for recovering objective A1 in Section 2.2 because $\hat{\theta}_i \to \theta$ implies $|\hat{v}_i(t) - v(t)| \to 0$. In this section, we restrict our attention to the group agreement problem as a special case of the adaptive design and show that the parameter convergence is achieved. We note, however, that the convergence to the desired target set (objective A2) is guaranteed by Theorem 4.1 even without parameter convergence.

We assume that $P_k(z_k)$'s are positive definite and radially unbounded functions on $\mathscr{G}_k = \mathbb{R}^p$ such that (2.46) is satisfied and thus, Property 2.1 holds. We further assume that the passive feedback block $\mathscr{H}_i$ is in the control affine form

$$\dot{\xi}_i = f_i(\xi_i) + g_i(\xi_i)u_i \tag{4.20}$$

$$y_i = h_i(\xi_i) \tag{4.21}$$

where

$$h_i(0) = 0, \qquad f_i(0) = 0 \tag{4.22}$$

and that the regressor $\Phi(t)$ in (4.4) is *persistently exciting* (PE), which means that for all $t_o \geq 0$,

$$\int_{t_o}^{t_o+\delta} \Phi(t)\Phi(t)^T dt \geq \alpha I \tag{4.23}$$

with some constants $\delta > 0$ and $\alpha > 0$ that do not depend on t_o. This PE condition ensures the information richness of the time-varying signal $\Phi(t)$ throughout time, and guarantees parameter convergence:

Theorem 4.2. *In Theorem 4.1, suppose that the desired sets are* $\mathscr{A}_k = \{0\}$, *and that the passive feedback block is of the form (4.20)-(4.22). If* $\Phi(t)$ *satisfies the PE condition (4.23), then the origin of* $(z, \xi, \tilde{\theta})$ *is globally uniformly asymptotically stable. In particular,* $\hat{\theta}_i \to \theta$, $i = 2, \cdots, N$ *as* $t \to \infty$. □

Proof. To prove parameter convergence in this case, we use the Nested Matrosov Theorem reviewed in Appendix B.5. The first auxiliary function V_1 is the same as the Lyapunov function V in (4.17), which yields the negative semidefinite derivative in (4.18) and thus guarantees uniform global stability, that is,

$$\dot{V}_1 = \dot{V} \leq \sum_{i=1}^{N} -W_i(\xi_i) := Y_1 \leq 0. \tag{4.24}$$

The second auxiliary function is

$$V_2 = z^T (D \otimes I_p)^+ \Gamma y \tag{4.25}$$

where $(D \otimes I_p)^+$ denotes the pseudoinverse of $D \otimes I_p$ and

$$\Gamma = \mathrm{diag}\{(L_{g_1} h_1(0))^{-1}, \cdots, (L_{g_N} h_N(0))^{-1}\}. \tag{4.26}$$

In particular $L_{g_i} h_i(0) := \frac{\partial h_i(\xi_i)}{\partial \xi_i}\Big|_{\xi_i=0} g_i(0)$ is nonsingular and thus invertible because of the passivity of the ξ_i-subsystems in (4.20) and because of Proposition B.1 in Appendix B.4. The derivative of V_2 yields

$$\dot{V}_2 = z^T (D \otimes I_p)^+ \Gamma \dot{y} + \dot{z}^T (D \otimes I_p)^+ \Gamma y := Y_2 \tag{4.27}$$

where we claim that

$$Y_1 = 0 \quad \Rightarrow \quad Y_2 \leq 0. \tag{4.28}$$

To see this, note that $Y_1 = 0$ implies $\xi = 0$ and it follows from (4.22) that $y = 0$, which means that the second term in $\dot{V}_2$ vanishes. Because $\dot{y}_i = L_{g_i} h_i(0) u_i$ when $\xi = 0$, Y_2 becomes

$$Y_2 = z^T (D \otimes I_p)^+ u. \tag{4.29}$$

Substituting (2.26) and $z^T = x^T (D \otimes I_p)$ from (2.6), we obtain

$$\begin{aligned} Y_2 &= -x^T (D \otimes I_p)(D \otimes I_p)^+ (D \otimes I_p) \psi(z) \\ &= -x^T (D \otimes I_p) \psi(z) \end{aligned}$$

$$= -z^T \psi(z) \leq 0. \tag{4.30}$$

Next we introduce the auxiliary function

$$V_3 = -((D^T \otimes I_p)\tilde{v})^T z \tag{4.31}$$

where $\tilde{v}$ is defined in (4.14). Its derivative is

$$\dot{V}_3 = -((D^T \otimes I_p)\tilde{v})^T \dot{z} - ((D^T \otimes I_p)\dot{\tilde{v}})^T z := Y_3 \tag{4.32}$$

and we claim

$$Y_1 = 0,\ Y_2 = 0 \ \Rightarrow\ Y_3 = -\left\{(D^T \otimes I_p)\tilde{v}\right\}^T \left\{(D^T \otimes I_p)\tilde{v}\right\} \leq 0. \tag{4.33}$$

To show (4.33), we first note that $Y_2 = 0$ implies that $z^T \psi(z) = 0$ and thus $z = 0$ due to (2.46), which means that the second term in (4.32) vanishes. It follows from $Y_1 = 0$ that $\xi = 0$ and hence y is zero from (4.22). Therefore, $\dot{z}$ in (3.55) becomes $(D^T \otimes I_p)\tilde{v}$, which proves (4.33).

Finally, we define the auxiliary function

$$V_4 = -\tilde{\theta}^T S(t)\tilde{\theta} \tag{4.34}$$

$$S(t) := \int_t^\infty e^{(t-\tau)} F(\tau)F(\tau)^T d\tau \quad F(t) := I_N \otimes \Phi(t) \otimes I_p \tag{4.35}$$

where

$$S(t) \geq \int_t^{t+\delta} e^{(t-\tau)} F(\tau)F(\tau)^T d\tau \geq \alpha e^{-\delta} I \tag{4.36}$$

because of the PE property of $\Phi(t)$. Note that

$$\begin{aligned}\dot{S}(t) &= e^t \int_t^\infty e^{-\tau} F(\tau)F(\tau)^T d\tau + e^t \frac{d}{dt}\left\{\int_t^\infty e^{-\tau} F(\tau)F(\tau)^T d\tau\right\} \\ &= S(t) - F(t)F(t)^T.\end{aligned} \tag{4.37}$$

From (4.37), we obtain

$$\dot{V}_4 \leq -\tilde{\theta}^T S(t)\tilde{\theta} + \tilde{v}^T \tilde{v} - 2\tilde{\theta}^T S(t)\dot{\tilde{\theta}} := Y_4 \tag{4.38}$$

and claim

$$Y_2 = 0,\ Y_3 = 0 \ \Rightarrow\ Y_4 = -\alpha e^{-\delta}|\tilde{\theta}|^2 \leq 0 \tag{4.39}$$

because the second and third terms in (4.38) vanish when $Y_2 = 0$ and $Y_3 = 0$. Indeed, $Y_3 = 0$ leads to $(D^T \otimes I_p)\tilde{v}(t) = 0$, which indicates that $\tilde{v}(t)$ lies in $\mathcal{N}(D^T \otimes I_p)$. Recall that $\mathcal{N}(D^T \otimes I_p) = 1_N \otimes c$, $c \in \mathbb{R}^p$ and $\tilde{v}_1 \equiv 0_p$. Therefore, it follows that $\tilde{v}(t) = 0$, which means that the second term in Y_4 (4.38) is zero. Likewise, from (2.46), $Y_2 = 0$ results in $z = 0$, which means z belongs to the desired set $\mathcal{A}$ and thus $\psi(z)$ and u are zero. It follows that $\dot{\tilde{\theta}} = 0$ from (4.11), which shows that the third term in Y_4 vanishes.

Because $Y_i = 0$, $i = 1,2,3,4$, imply $(\xi, z, \tilde{\theta}) = 0$ and we conclude from the Nested Matrosov Theorem in Appendix B.5 that the origin is globally uniformly asymptotically stable. Note that $\tilde{\theta} \to 0$ implies $\hat{\theta}_i \to \theta$, $i = 2, \cdots, N$. □

4.4 The Augmented Design

Like the design in Section 3.3, the basic adaptive design in Section 4.2 ensures tracking of reference velocity (objective A1) only in special cases, such as the agreement problem studied in the previous section. To guarantee objective A1 when the basic adaptive design fails, we employ the augmented design (3.68) in the update law (4.8). The augmented design recovers the stability result of Theorem 2.1 and achieves tracking of the reference velocity as well.

Theorem 4.3. *Consider the coordination laws in (3.23), (3.24) and (4.8), where $v(t)$ is parameterized as (4.1) in which $\phi^j(t)$, $\dot{\phi}^j(t)$, $j = 1, \cdots, r$ are continuous and uniformly bounded. With u_i defined in (3.68), and $\mathscr{H}_i$, $i = 1, \cdots, N$, and ψ_k, $k = 1, \cdots, \ell$ designed as in (2.11)-(2.15) and (2.27)-(2.31), all trajectories $(z(t), \xi(t), \hat{\theta}(t))$ starting in $\mathscr{G} \times \mathbb{R}^{pr(N-1)}$ are bounded and converge to the set*

$$\mathscr{E}_p^* = \left\{(z, \xi, \hat{\theta}) | \xi = 0, (D \otimes I_p)\psi(z) = 0, z \in \mathscr{R}(D^T \otimes I_p), \hat{v}_i(t) = v(t)\right\}, \quad (4.40)$$

where $\hat{v}_i(t)$ is defined in (4.4) and $\mathscr{G}$ is as in (2.37). □

The closed-loop structure of the augmented adaptive design is given in Fig. 4.2. Recall from Property 1.4 that $\dot{x}^T(L^v \otimes I_p)\dot{x} = \frac{1}{2}\dot{x}^T(L^v_{\text{sym}} \otimes I_p)\dot{x}$, which is nonnegative since G^v is strongly connected and balanced. Thus, the static block L^v_{sym} in Fig. 4.2 is passive and the feedforward path from $\dot{x}$ to $-u$ is also passive. As the passivity of the feedback path is already established in Theorem 4.1, we conclude the closed-loop stability of the system in Fig. 4.2.

Proof. Note that (3.68) can be rewritten in the compact form (3.69). To prove the stability of the closed-loop system described by the adaptive design (2.11), (3.55), (3.69) and (4.11), we take the same Lyapunov function as in (4.17) and from (2.40), (3.70), (4.16) and (3.69), compute its time derivative as

$$\dot{V} = -\sum_{i \in \mathscr{I}} W_i(\xi_i) - \sum_{i \notin \mathscr{I}} u_i^T y_i - (y + \tilde{v})^T (L^v \otimes I_p)(y + \tilde{v}) \leq 0 \quad (4.41)$$

which implies stability and boundedness of $(z(t), \xi(t), \tilde{\theta}(t))$. Using Theorem B.5, we further conclude that $\xi_i \to 0$, $\forall i \in \mathscr{I}$, $u_i \to 0$, $\forall i \notin \mathscr{I}$ and $(y + \tilde{v})^T (L^v \otimes I_p)(y + \tilde{v}) \to 0$. For dynamic block $\mathscr{H}_i$, it follows from $\xi_i \to 0$ and (2.11) that $y_i \to 0$. For static block $\mathscr{H}_i$, $u_i \to 0$ implies $y_i = h_i(u_i) \to 0$. Thus, $y \to 0$. Recall from (1.20) that $(y + \tilde{v})^T (L^v \otimes I_p)(y + \tilde{v})$ is zero only when $(y + \tilde{v})^T (L^v_{\text{sym}})(y + \tilde{v})$ is zero. Since the graph G^v is strongly connected, the graph corresponding to L^v_{sym} is connected. Therefore, $(y + \tilde{v})^T (L^v \otimes I_p)(y + \tilde{v}) = 0$ implies $y + \tilde{v} = 1_N \otimes c$, where $c \in \mathbb{R}^p$. We

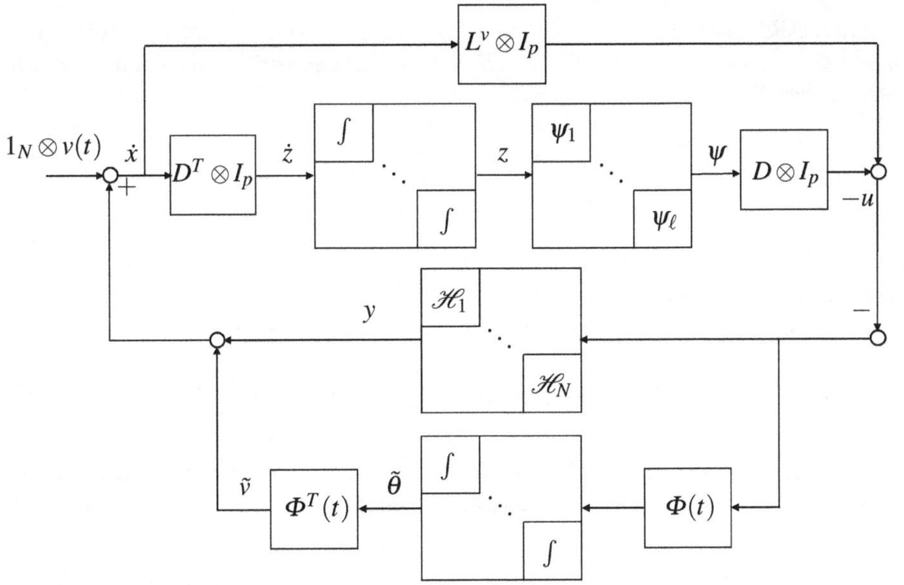

Fig. 4.2 The closed-loop structure of the augmented adaptive design. The property of L^v in (1.20) renders the passivity of the top loop. The closed-loop stability follows from the interconnection of two passive feedforward paths and two passive feedback paths.

conclude from $\tilde{v}_1 \equiv 0$ and $y_1 \to 0$ that $|y+\tilde{v}| \to 0$. Since $y \to 0$, it follows that $\tilde{v} \to 0$, which implies from (4.14) that tracking of $v(t)$ is achieved.

We next show $u \to 0$. To this end we note that

$$\ddot{\xi}_i = \frac{\partial f_i}{\partial u_i}\dot{u}_i + \frac{\partial f_i}{\partial \xi_i}\dot{\xi}_i \tag{4.42}$$

is continuous and uniformly bounded because $\dot{u}$ and $\dot{\xi}$ are continuous functions of the bounded signals $(z(t), \xi(t), \tilde{\theta}(t), \Phi(t), \dot{\Phi}(t))$ and because $f_i(\cdot,\cdot)$ is C^1. Since $\xi_i \to 0$ and $\ddot{\xi}_i$ is continuous and bounded, it follows from Theorem B.4 that $\dot{\xi}_i \to 0$, which, from (2.11) and (2.12), guarantees $u_i \to 0$. Since $|y+\tilde{v}| \to 0$, we conclude from (3.69) that $(D \otimes I_p)\psi(z) \to 0$. □

The main difference of Theorem 4.3 from Theorem 4.2 is that it achieves reference velocity tracking directly while Theorem 4.2 establishes tracking by first achieving parameter convergence. Parameter convergence is sufficient but not necessary for velocity tracking (compare (4.1) and (4.4)). We next show that velocity tracking implies parameter convergence $\hat{\theta}_i \to \theta$ when the regressor $\Phi(t)$ satisfies the PE condition in (4.23). We need the following lemma.

Lemma 4.1. *Let*

$$\dot{X} = f(X,t), \tag{4.43}$$

where $X \in \mathbb{R}^n$ and $f(X,t): \mathbb{R}^n \times \mathbb{R}_{\geq 0} \to \mathbb{R}^n$. If all trajectories $X(t)$ satisfy $f(X(t),t) \to 0$ and $\Omega(t)^T X(t) \to 0$, where $\Omega(t) \in \mathbb{R}^n$ is bounded and satisfies the PE property in (4.23), then $X(t) \to 0$. □

Proof. We rewrite (4.43) as

$$\dot{X} = -\Omega(t)\Omega(t)^T X + \zeta(t) \tag{4.44}$$

where $\zeta(t) := \Omega(t)\Omega(t)^T X + f(X,t)$, and note that $\zeta(t) \to 0$ since $\Omega(t)^T X$ and $f(X,t)$ both converge to zero and since $\Omega(t)$ is bounded. Solving for X from the linear time-varying model (4.44), we obtain

$$X(t) = \Xi(t,t_0)X(t_0) + \int_{t_0}^{t} \Xi(t,\tau)\zeta(\tau)d\tau \tag{4.45}$$

where $\Xi(t,t_0)$ is the state transition matrix. Because $\Omega(t)$ is PE and because $\zeta(t) \to 0$ as $t \to \infty$, it follows from standard results in adaptive control (e.g., [62, 136]) that $X(t) \to 0$. □

We now combine Theorem 4.3 and Lemma 4.1 to prove parameter convergence:

Corollary 4.1. *Suppose all conditions of Theorem 4.3 hold. If, in addition, $\Phi(t)$ satisfies (4.23), then $\hat{\theta}_i \to \theta$.* □

Proof. We establish $\hat{\theta}_i \to \theta$ by using the PE property (4.23) and Lemma 4.1 to prove that $|\tilde{v}| \to 0$ implies $\tilde{\theta}_i \to 0$, that is $\hat{\theta}_i \to \theta$.

We note from Theorem 4.3 that

$$\tilde{v} = (I_N \otimes \Phi^T(t) \otimes I_p)\tilde{\theta} \to 0 \tag{4.46}$$

and that

$$\dot{\hat{\theta}}_i = \Lambda_i(\Phi(t) \otimes I_p)u_i \to 0 \tag{4.47}$$

since $u_i \to 0$. Because the signal $\Phi^T(t)$ is PE, it follows from Lemma 4.1 that $\tilde{\theta}_i \to 0$, which proves the parameter convergence $\hat{\theta}_i \to \theta$. □

Example 4.2. To illustrate the parameter convergence, we simulate the example in Section 3.5.1. We take

$$v(t) = ([\sin(t)\ \cos(t)] \otimes I_2)\begin{bmatrix}\theta^1 \\ \theta^2\end{bmatrix} \tag{4.48}$$

where $\theta^1 = [-\frac{\sqrt{3}}{3}\ 0]^T$ and $\theta^2 = [0\ \frac{\sqrt{3}}{3}]^T$. This $v(t)$ is the same as in Section 3.4.2. The estimate $\hat{v}_i(t)$ in (3.64) is obtained from (4.4) with $\Phi(t) = [\sin(t)\ \cos(t)]^T$ and $\hat{\theta}_i = [(\hat{\theta}_i^1)^T\ (\hat{\theta}_i^2)^T]^T$ updated by (4.8).

The initial conditions of $x_i(0)$, $i = 1,2,3$, $\xi_1(0)$, and $\hat{\xi}_i(0)$, $i = 2,3$ are the same as in Section 3.4.2. The initial estimates are set to $\hat{\theta}_2(0) = [\frac{\sqrt{3}}{6}\ -\frac{1}{2}\ -\frac{1}{2}\ -\frac{\sqrt{3}}{6}]^T$ and $\hat{\theta}_3(0) = [\frac{\sqrt{3}}{6}\ \frac{1}{2}\ \frac{1}{2}\ -\frac{\sqrt{3}}{6}]^T$ such that the group exhibits the same motion as in Fig. 3.4(b) if (2.23) is used in (4.8).

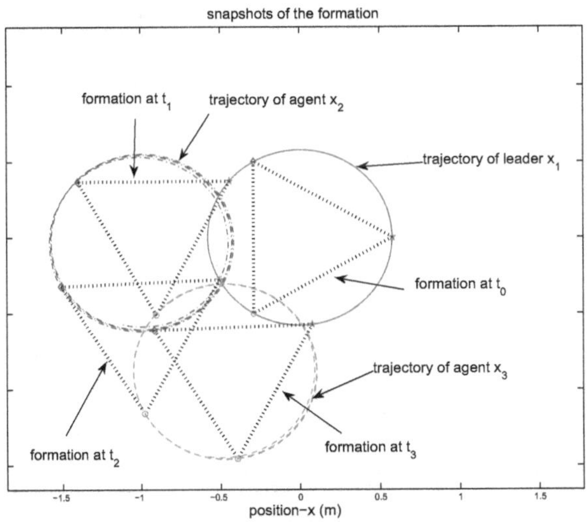

Fig. 4.3 The augmented adaptive design recovers the convergence properties of the nonadaptive design. [13]. Reprinted with Permission. Copyright Elsevier 2009.

When the augmented feedback (3.68) is employed in (4.8), Fig. 4.3 shows the snapshots of the formation. The group now exhibits a translational motion with x_1 circling around the origin, which means that the nonadaptive results are fully recovered. In addition, because $\Phi(t)$ is PE, parameter convergence is achieved as shown in Fig. 4.4. In this simulation, the graphs G and G^v are chosen the same as in Section 3.5.1. □

4.5 Application to Gradient Climbing in Formation

In this section, we apply the adaptive design result to a gradient climbing problem, where the group leader performs extremum seeking for the field minima or maxima, while the other agents maintain a desired formation with respect to the leader. Keeping a group formation during the gradient climbing may be desirable for reliable inter-vehicle communication/sensing, drag reduction, safety in adversarial environments, etc.

To achieve gradient climbing in a field distribution, the leader takes a discrete-time, optimization based extremum seeking approach. This extremum-seeking approach, illustrated in Fig. 4.5, generates finite-difference approximations for the gra-

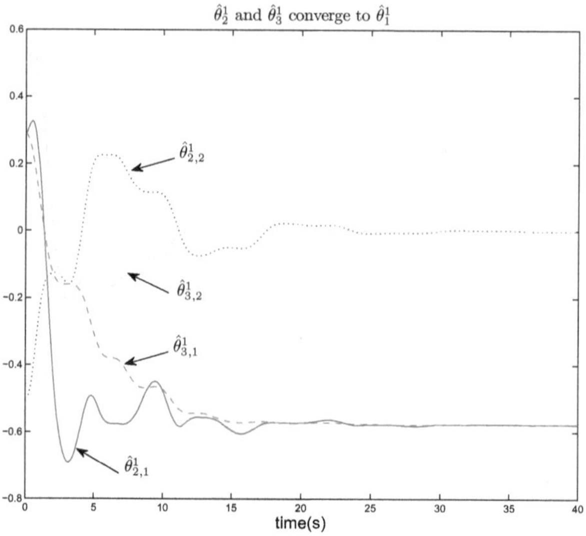

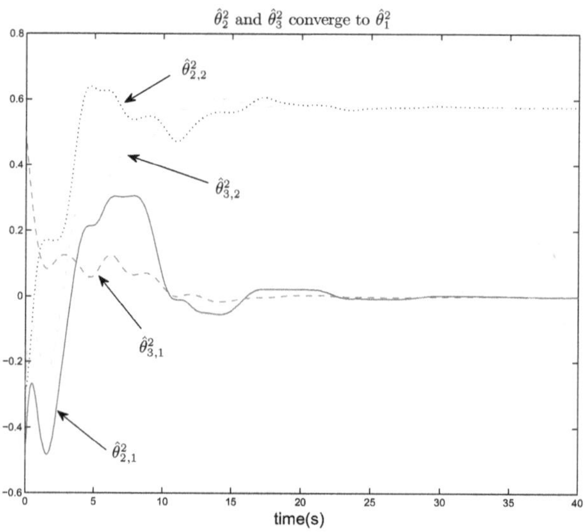

Fig. 4.4 Parameter convergence with the augmented adaptive design. $\hat{\theta}_2^1$ and $\hat{\theta}_3^1$ converge to $\theta^1 = [-\frac{\sqrt{3}}{3}\ 0]^T$ while $\hat{\theta}_2^2$ and $\hat{\theta}_3^2$ converge to $\theta^2 = [0\ \frac{\sqrt{3}}{3}]^T$. [13]. Reprinted with Permission. Copyright Elsevier 2009.

dient and the Hessian of the field, by "dithering" sensor positions. The advantage of this local approximation is that only the leader needs sensing capabilities, and communication of sensed variables and geographic proximity of sensors are not necessary for generating approximate gradients.

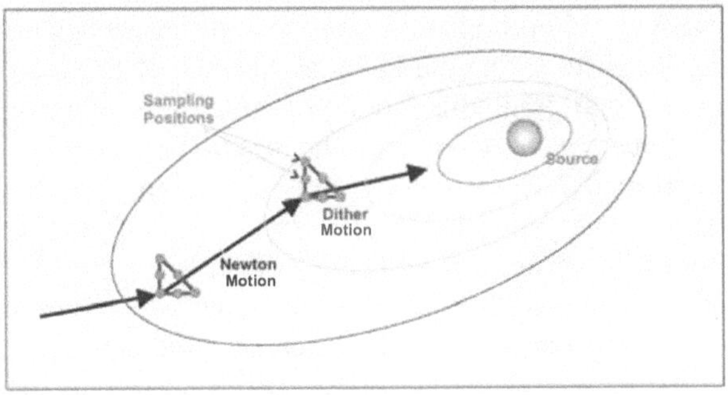

Fig. 4.5 Gradient climbing by extremum seeking. Arrows represent the Newton motion, while triangular paths are the dither motion with the samples taken at positions marked by dots. The dither motion has three segments: Along horizontal axis from left to right, along the diagonal from right to left and along vertical axis from top to bottom. The directions of these three segments are denoted by $[1,0]$, $[-1,1]$, $[0,-1]$.

After the dither motion, the leader calculates a Newton direction towards the field extremum. Thus, the group reference velocity $v(t)$ is determined autonomously by the leader, in the form of segments $v_k(t)$, $t \in [t_k, t_{k+1}]$, that are updated in every iteration k according to the next Newton direction. Since $v(t)$ is not available to the other agents, they need to estimate this $v(t)$ information to achieve a successful gradient climbing in the desired formation.

We let the leader parameterize its reference velocity as in (4.1) and apply the *basic* adaptive design in this chapter to ensure a desired group formation. During the dither motion of the leader, the other agents may turn off the velocity adaptation design so that they do not respond to the dither motion of the leader. Even if the adaptation is not turned off, we show with simulation results that if the Newton motion lasts sufficiently long, the followers respond only to the Newton motion while filtering out the dither component.

4.5.1 Reference Velocity Assignment by the Leader

In this section, we present the extremum seeking scheme performed by the leader. The analysis of the motion of the group will be pursued in Section 4.5.2. The goal in extremum-seeking based gradient climbing is to search for and move towards the maximum of a field distribution with an unknown functional form. The leader has access only to the scalar field measurements, and constructs the approximate gradient and Hessian information of the field by finite-difference methods to compute a Newton direction. It then assigns an appropriate velocity along the computed Newton direction. It is important to note that this scheme locates the maxima without position measurements.

We first review basic optimization tools that are instrumental in the extremum seeking design. We assume that the field has a spatial distribution characterized by a twice continuously differentiable function $F(x) : \mathbb{R}^2 \to \mathbb{R}$ that has a unique maximum at $x = x^*$. Note that we restrict our attention to fields only in $\mathbb{R}^2$, however, the results can be extended to $\mathbb{R}^3$ as well by employing appropriate finite-difference approximations. Also note that if the function $F(x)$ has multiple maxima, then the results can be modified to prove regional convergence to the local maximum. Because only field measurements are available to the leader, we approximate the gradient and Hessian of $F(x)$ by one-sided finite-difference gradient, G_k,

$$\nabla F(x_k) \approx G_k[i] := \frac{F(x_k + h_k e_i) - F(x_k)}{h_k} \tag{4.49}$$

and Hessian, H_k,

$$\begin{aligned}\nabla^2 F(x_k) \approx H_k[i,j] := \frac{1}{h_k^2}\Big[&F(x_k) + F(x_k + h_k e_i + h_k e_j) \\ &- F(x_k + h_k e_i) - F(x_k + h_k e_j)\Big]\end{aligned} \tag{4.50}$$

where h_k denotes the finite-difference "dither" size, and e_i is the ith unit vector. For an easier implementation, steepest descent may be preferable over Newton's Method; however, it is slower and does not provide a concrete convergence proof with nonvanishing step-size. We denote by $\mathbf{B}(\bar{x}, a)$ the ball of radius a centered at $\bar{x}$, i.e., $\mathbf{B}(\bar{x}, a) := \{x|\ |x - \bar{x}| \le a\}$. The lemma below states that for sufficiently small dither size h_k, and for small initial error $|x_0 - x^*|$, finite-difference based Newton's Method locally converges to an $O(h)$-neighborhood of x^*. The proof follows from standard arguments in unconstrained optimization theory, and is given in Appendix A.3.

Lemma 4.2. *Let $F(x) : \mathbb{R}^2 \to \mathbb{R}$ be twice continuously differentiable in an open convex set $\mathscr{D} \in \mathbb{R}^2$. Assume there exists a unique $x^* \in \mathbb{R}^2$, $r > 0$ and $\beta > 0$ such that $\mathbf{B}(x^*, r) \in \mathscr{D}$, $\nabla F(x^*) = 0$, $\nabla^2 F(x^*)^{-1}$ exists with $||\nabla^2 F(x^*)^{-1}|| \le \beta$, and $\nabla F(x)$ and $\nabla^2 F(x)$ are Lipschitz continuous. Then there exist $\varepsilon > 0$ and $\bar{h} > 0$, such that for all initial conditions $x_0 \in \mathbf{B}(x^*, \varepsilon)$, and dither size $h_k < \bar{h}$ the sequence $\{x_k\}_{k=0}^{k=\infty}$ generated by*

$$x_{k+1} = x_k + H_k^{-1} G_k, \quad k = 0, 1, \cdots \tag{4.51}$$

where G_k and H_k are as in (4.49)-(4.50) converges to an $O(\bar{h})$ neighborhood of x^* q-linearly. □

We next introduce the Newton's Method-based gradient climbing scheme that the leader implements to locate the maximum of a field. We consider the agent model in (2.16) and the control design (2.17). We assume that the leader, say agent 1, does not receive external feedback u_i from other members of the group, hence $u_1 \equiv 0$. Recall that (2.16) and (2.17) can be transformed to (2.19) and (2.20). We then assume $\xi_1(0) = 0$, which implies from (2.19) and (2.20) that

$$\dot{x}_1 = v(t). \tag{4.52}$$

Note that if $\xi_1(0) \neq 0$, we can apply the velocity input $\bar{v}(t) = -\xi_1 + v(t)$ to the leader, and recover (4.52).

We use Newton's Method to determine the next position for the leader, and set the reference velocity $v(t)$ to steer the leader to that position. As illustrated in Fig. 4.5, in the kth extremum-seeking iteration, the leader first moves in $[1,\ 0]$, $[-1,\ 1]$, and $[0,\ -1]$ directions rapidly to take samples of the field $F(x)$ and computes the approximate gradient G_k and the Hessian H_k as in (4.49)-(4.50), and then moves in the approximate Newton direction $l_k = H_k^{-1} G_k$, and arrives at $x_{k+1} = x_k + l_k$.

To prepare for an adaptive reference velocity estimation by the followers, we parameterize $v(t)$ in a form similar to (4.1). For each motion segment we let the reference velocity have a fixed sinusoidal amplitude profile, with endpoints at zero, and change its direction between successive segments. We denote by $v_{[i,j]}$ and v_N the dither velocity in the $[i,j]$ direction, where $[i,j] \in \{[1,0],[-1,1],[0,-1]\}$, and the Newton velocity in l_k direction, respectively. Let t_d be the duration of each dither motion segment, and T be that of the Newton motion. Therefore one iteration of the extremum seeking scheme takes $\Delta := 3t_d + T$ seconds. During each extremum seeking iteration, the leader switches its velocity as

$$\dot{x}_1 = v(t) := \begin{cases} v_{[1,0]}(t), & \text{if } t_k \leq t < t_k + t_d, \\ v_{[-1,1]}(t), & \text{if } t_k + t_d \leq t < t_k + 2t_d, \\ v_{[0,-1]}(t), & \text{if } t_k + 2t_d \leq t < t_k + 3t_d, \\ v_N(t), & \text{if } t_k + 3t_d \leq t < t_{k+1}, \end{cases} \tag{4.53}$$

where

$$t_k := k\Delta, \quad k = 0,1,2,\cdots, \tag{4.54}$$

and $v_{[1,0]}$, $v_{[-1,1]}$, $v_{[0,-1]}$ and v_N are defined as:

$$v_{[1,0]}(t) := \frac{2h_k}{t_d}\begin{bmatrix}1\\0\end{bmatrix}\left(1-\cos(\frac{2\pi}{t_d}(t-t_k))\right) \tag{4.55}$$

$$v_{[-1,1]}(t) := \frac{2h_k}{t_d}\begin{bmatrix}-1\\1\end{bmatrix}\left(1-\cos(\frac{2\pi}{t_d}(t-t_k-t_d))\right) \tag{4.56}$$

$$v_{[0,-1]}(t) := \frac{2h_k}{t_d}\begin{bmatrix}0\\-1\end{bmatrix}\left(1-\cos(\frac{2\pi}{t_d}(t-t_k-2t_d))\right) \tag{4.57}$$

$$v_N(t) := \frac{l_k}{T}\left(1-\cos(\frac{2\pi}{T}(t-t_k-3t_d))\right). \tag{4.58}$$

The reference velocity $v(t)$ in (4.53) and its derivative $\dot{v}(t)$ are continuous, and $(v(t),\dot{v}(t))|_{t\in\{t_k+nt_d,\ t_{k+1}\}} = (0,0)$, $n = 0,1,2,3$. Note that other continuous velocity profiles that vanish at $t \in \{t_k + nt_d,\ t_{k+1}\}$, $n = 0,1,2,3$, along with their derivatives, are also applicable. The velocities in (4.55)-(4.58), when switched according to (4.53), achieve one iteration of extremum-seeing motion by driving the leader first to the appropriate "dither" positions and then to the next "Newton" position

x_{k+1}. Theorem 4.4 below proves that the extremum seeking scheme converges to an $O(\bar{h})$ neighborhood of the maximum x^*, when $h_k \leq \bar{h}$ is as in Lemma 4.2, and $|x(0)-x^*|$ is sufficiently small.

Theorem 4.4. *Let the field distribution $F(x)$ be twice continuously differentiable with a unique maximum at position $x = x^* \in \mathbb{R}^2$. Suppose the assumptions in Lemma 4.2 hold and $\bar{h}$ be as defined therein. Then the Newton-based extremum seeking scheme applied to the vehicle model in (4.53) with velocity profiles (4.55)-(4.58) drives the vehicle to the $O(\bar{h})$ neighborhood of x^*, provided that $h_k \leq \bar{h}$ and $|x(0) - x^*|$ is sufficiently small.* □

Proof. We show that the reference velocity profiles given in (4.53) first drive the leader in the appropriate dither directions, and then along the Newton direction. Consider $v_{[1,0]}$ which drives the leader in horizontal position, i.e., along the vector $[1,0]$. At time t_k, let the position of the leader be $x_1(t_k) = [x_1^1(t_k), x_1^2(t_k)]^T \in \mathbb{R}^2$. Then at time $t_k + t_d/2$ its position is:

$$\begin{aligned}
x_1(t_k + \frac{t_d}{2}) &= x_1(t_k) + \int_{t_k}^{t_k+t_d/2} v_{[1,0]}(t)dt \\
&= x_1(t_k) + \frac{2h_k}{t_d}\begin{bmatrix}1\\0\end{bmatrix}\int_{t_k}^{t_k+t_d/2}(1-\cos(\frac{2\pi}{t_d}(t-t_k)))dt \\
&= x_1(t_k) + \frac{2h_k}{t_d}\begin{bmatrix}1\\0\end{bmatrix}\left[t - \frac{t_d}{2\pi}\sin(\frac{2\pi}{t_d}(t-t_k))\right]\Big|_{t_k}^{t_k+t_d/2} \\
&= x_1(t_k) + h_k\begin{bmatrix}1\\0\end{bmatrix} = \begin{bmatrix}x^1(t_k)+h_k\\ x^2(t_k)\end{bmatrix}.
\end{aligned} \tag{4.59}$$

Likewise,

$$\begin{aligned}
x_1(t_k + t_d) &= x_1(t_k + t_d/2) + \int_{t_k+t_d/2}^{t_k+t_d} v_{[1,0]}(t)dt \\
&= x_1(t_k + t_d/2) + h_k\begin{bmatrix}1\\0\end{bmatrix} = \begin{bmatrix}x^1(t_k)+2h_k\\ x^2(t_k)\end{bmatrix}.
\end{aligned} \tag{4.60}$$

Similar calculations show that $v_{[-1,1]}$ and $v_{[0,-1]}$ achieve the desired dither motions as well. Note that after the third dither motion $v_{[0,-1]}$ the leader will be back at position $x_1(t_k + 3t_d) = x_1(t_k)$. Then, applying the "Newton" velocity v_N after this point for T seconds drives the leader to

$$\begin{aligned}
x_1(t_k + \Delta) &= x_1(t_k) + l_k\frac{1}{T}\int_{t_k+3t_d}^{t_k+3t_d+T}(1-\cos(\frac{2\pi}{T}(t-t_k-3t_d)))dt \\
&= x_1(t_k) + l_k = x_1(t_{k+1}).
\end{aligned} \tag{4.61}$$

Therefore, by switching the velocities as in (4.53) the leader visits all dither positions and moves to the next Newton position. The convergence result follows from Lemma 4.2. □

4.5.2 Gradient Climbing in Formation

We have shown that using the switching strategy in (4.53) with the reference velocity $v(t)$ parameterized as in (4.55)-(4.58), the leader locates the extrema of the field. We next investigate how to design the motion of the other agents to achieve gradient climbing in a desired formation.

As discussed in Section 2.6, we may pursue the position-based or distance-based formation control formulation. In either formulation, we note that the other agents do not have the knowledge of the reference velocity $v(t)$ which changes after each iteration of extremum seeking. Therefore, the adaptive designs in Chapter 3 and this chapter can be applied to estimate the $v(t)$ information. Since (4.55)-(4.58) are already parameterized as a product of a vector and a time-varying basis function, we will take the parameterization approach in this chapter. Then the dynamics of agent i, $i = 2, \cdots, N$ are given by the basic adaptive design (3.24), (2.11), (4.4), and (4.8), where we assume that u_i has already been designed according to the position-based or distance-based formation control formulation. Following (4.53), we obtain the basis function $\Phi(t) \in \mathbb{R}$ and the constant parameter $\theta \in \mathbb{R}^2$ in (4.1) as

$$\Phi(t) := \begin{cases} \left(1 - \cos(\frac{2\pi}{t_d}(t - t_k))\right), & \text{if } t_k \leq t < t_k + t_d, \\ \left(1 - \cos(\frac{2\pi}{t_d}(t - t_k - t_d))\right), & \text{if } t_k + t_d \leq t < t_k + 2t_d, \\ \left(1 - \cos(\frac{2\pi}{t_d}(t - t_k - 2t_d))\right), & \text{if } t_k + 2t_d \leq t < t_k + 3t_d, \\ \left(1 - \cos(\frac{2\pi}{T}(t - t_k - 3t_d))\right), & \text{if } t_k + 3t_d \leq t < t_{k+1}, \end{cases} \tag{4.62}$$

and

$$\theta := \begin{cases} \frac{2h_k}{t_d}\begin{bmatrix} 1 \\ 0 \end{bmatrix}, & \text{if } t_k \leq t < t_k + t_d, \\ \frac{2h_k}{t_d}\begin{bmatrix} -1 \\ 1 \end{bmatrix}, & \text{if } t_k + t_d \leq t < t_k + 2t_d, \\ \frac{2h_k}{t_d}\begin{bmatrix} 0 \\ -1 \end{bmatrix}, & \text{if } t_k + 2t_d \leq t < t_k + 3t_d, \\ \frac{l_k}{T}, & \text{if } t_k + 3t_d \leq t < t_{k+1}. \end{cases} \tag{4.63}$$

In each motion segment, agent i employs the basic adaptive design in Section 4.2 to estimate θ by $\hat{\theta}_i$ and reconstruct the desired formation.

If t_d and T are sufficiently large, the result in Theorem 4.1 implies that the desired formation is ensured during each motion segment. This means that the agents will follow both the dither and the Newton motions of the leader. However, if only the leader has the sensing ability, it may be desired that the other agents respond only to the Newton motion. This can be achieved by simply turning off the adaptation during the dither motion periods. Even if adaptation is not turned off, the other agents detect only the Newton motion if T is sufficiently larger than t_d. To see this, we note from (4.53) that the average velocity of $v(t)$ within one extremum seeking

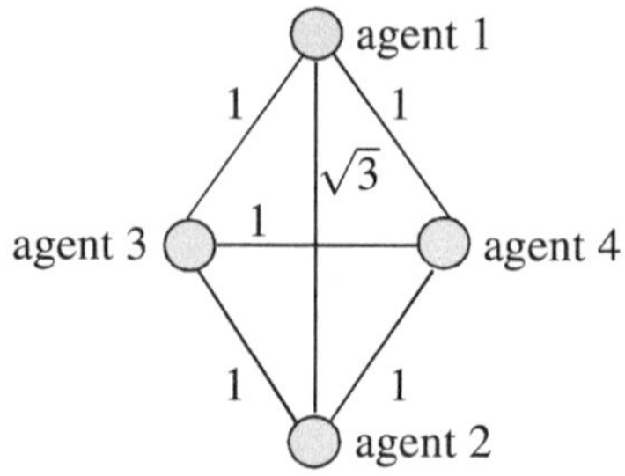

Fig. 4.6 The desired formation of four agents in the gradient climbing. The number indicates the desired length of each link.

iteration is given by

$$v_{av} := \frac{1}{\Delta}\int_{t_k}^{t_k+\Delta} v(t)dt = \frac{1}{\Delta} l_k. \tag{4.64}$$

Then if T is sufficiently large, $\Delta \approx T$ and thus $v_{av} \approx \frac{1}{T} l_k$, which is indeed the average of $v_N(t)$ in (4.58) in one Newton motion period. This implies that for sufficiently large T, the dither motion is averaged out. Thus, we can choose a sufficiently large T to ensure that the other agents follow only the Newton motion within each extremum-seeking iteration. In fact, for large T, one can further reveal a time-scale separation behavior in the group motion and show that the convergence to the desired formation is achieved in the fast time scale, while the Newton motion is performed in the slow time-scale. We refer interested readers to [18] for further details.

4.5.3 Simulation Results

We simulate the gradient climbing of four agents modeled by (2.16) with $m_i = 1$. We consider the distance-based formation control in Section 2.7.1. The desired formation is a rhombus formation shown in Fig. 4.6. Note from Example 2.5 that to ensure an unambiguous desired formation of four agents, we need to specify the desired relative distances between every two agents. According to Fig. 4.6, we define $z_1 = x_1 - x_2$ and set its desired distance as $d_1 = \sqrt{3}$. For the other z_k's, $k = 2, \cdots, 6$, their desired distances are $d_k = 1$. Given d_k's, the nonlinearity $\psi_k(z_k)$ can be designed according to (2.83)-(2.87). For the simulations we take $\sigma_1(s) = \ln(s/\sqrt{3})$ and $\sigma_k(s) = \ln(s)$, $k = 2, \cdots, 6$.

We let the field distribution be

$$F(x,y) = e^{-0.1e^{0.1x}(1.1x-5)^2 - 0.2e^{0.1y}(0.8y-4)^2},$$

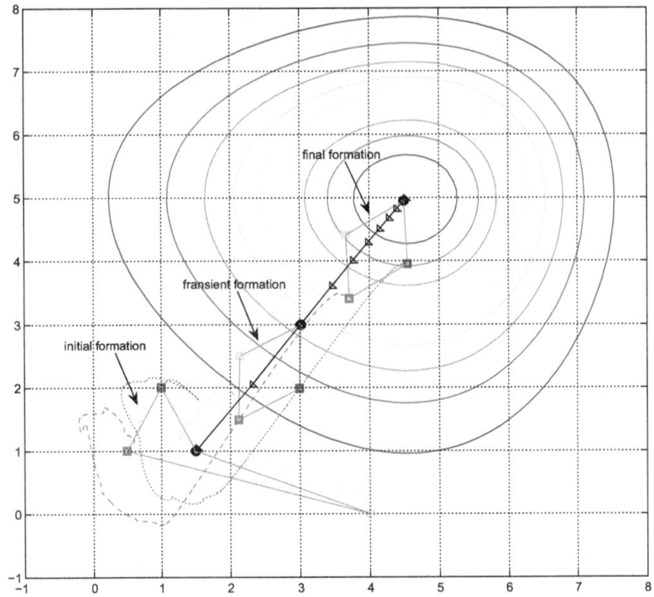

Fig. 4.7 Gradient climbing by Newton-based extremum seeking with $T = 18.5$ sec, $t_d = 0.5$ sec, and $h_k = 0.05$. Solid line represents the leader's trajectory, while dashed line, dash-dot line, and dots are the followers'. After an initial transient agents follow the leader's Newton motion in a rhombus formation, and average out the fast dither perturbations.

which has a global maximum at $x = [4.55, 5]^T$. We fix $\Delta = 20$ sec and $h_k = 0.05$, and pick $t_d = 0.5$ sec and $T = 18.5$ sec for the first simulation. We run the system (4.52) and (3.24), where the leader determines its velocity by extremum seeking as in (4.53) and (4.55)-(4.58) and the other agents estimate $v(t)$ by (4.4) with $\Phi(t)$ in (4.62). Fig. 4.7 shows that after an initial transient, agents follow the leader's Newton motion in a rhombus formation, and average out the fast dither perturbations, while the leader locates the maxima of the field. In the second simulation, we perform the dither motion at a slower speed with $t_d = 4$ sec, $T = 8$ sec. In this case, the agents in Fig. 4.8 fail to average out the dither motion, and follow a jittering trajectory.

4.6 Summary

In Chapters 3 and 4, we studied a group coordination problem where the reference velocity is available to only one agent while the others estimate this information with adaptive designs. We presented two approaches to the adaptive designs. The first approach assumes a constant or a periodic reference velocity while the second approach parameterizes the reference velocity as linear combinations of known time-varying basis functions with unknown coefficients. For each approach, we first

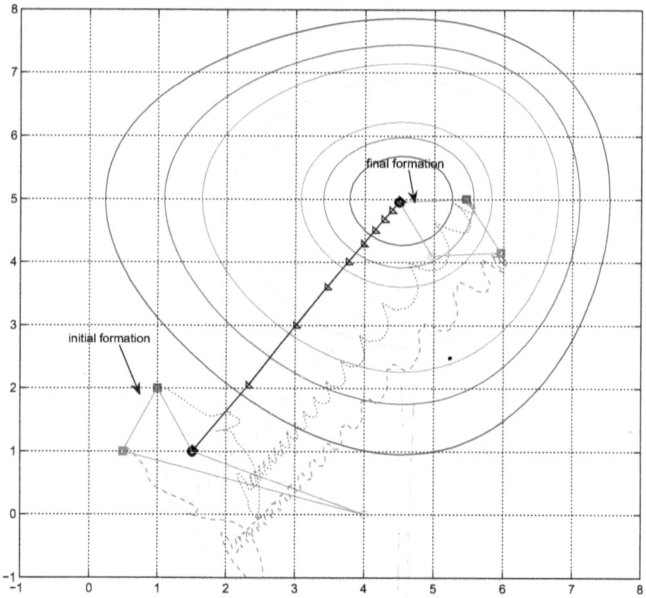

Fig. 4.8 Gradient climbing by Newton-based extremum seeking with $T = 8$ sec, $t_d = 4$ sec, and $h_k = 0.05$. Solid line represents the leader's trajectory, while dashed line, dash-dot line, and dots are the followers'. The agents fail to average out the dither motion, and follow a jittering trajectory.

proposed a basic adaptive design that guarantees objective A2. We showed that tracking of the reference velocity is recovered for some special cases including the agreement problem. We presented an example which shows that the estimates of the reference velocity may fail to converge to a time-varying reference velocity. For each approach, we then proposed an augmented adaptive redesign that employs relative velocity feedback in addition to relative position feedback and achieves tracking of the reference velocity.

We next applied the basic adaptive design in this chapter to an extremum seeking example, where the leader autonomously determines the Newton direction based on samples of a field distribution and parameterizes the group reference velocity according to the Newton direction. The other agents then estimate the reference velocity using the parameterization approach. In the simulation, we showed that if the Newton motion lasts sufficiently long within each extremum seeking period, the desired formation is reconstructed during gradient climbing.

4.7 Notes

- The extremum seeking results in this chapter are based on [18].

- The extremum seeking approach in this chapter relies on nonlinear optimization techniques to estimate the gradient in discrete time. An alternative approach in extremum seeking is to probe the system with sinusoidal inputs, and to make an online estimation of the gradient of the output relative to these inputs [151, 7].
- To enhance robustness to noise and input disturbance, existing modifications of adaptive design, such as σ-leakage modification [62], can be applied to the adaptive designs in Chapter 3 and this chapter.

Chapter 5
Attitude Coordination Without Inertial Frame Information

5.1 Introduction

The previous chapters modeled the agents as point robots and considered only position control. However, in numerous applications, the attitude of the agents plays an important role, which means that the agents must be modeled as rigid bodies. In this chapter, we study an attitude agreement problem where each agent has access only to the relative attitude with respect to its neighboring agents and to its angular velocity in its own body frame. An important application of attitude agreement is space interferometry, where spacecraft need to maintain the same attitude. It is important to note that we do not rely on inertial attitude measurements, which would require star tracker sensors that in practice are limited by their low-bandwidth. In contrast, relative attitude can be obtained by visual means and the angular velocity can be measured with a gyro.

To develop decentralized controllers for group alignment, we extend the passivity-based framework in Section 2.3 from the point-based robot formation control to attitude control problem. Because the inertial frame is not available to the agents, we represent the relative attitude in local body frames. The use of local body frames distinguishes the orientation coordination problem from the framework in Section 2.2, where the relative position is represented in the inertial frame.

We also draw on the earlier results on single rigid body attitude control [144], which exploit inherent passivity properties of rigid body dynamics and quaternion kinematics. In particular, we present a quaternion-based decentralized controller that locally asymptotically stabilizes the desired equilibria in which all the rigid bodies possess the same attitude, and rotate at the reference angular velocity. The resulting closed-loop structure is similar to the structure in Fig. 2.2. We further consider a situation similar to Chapters 3 and 4, where the reference angular velocity is available only to the leader. Following similar design techniques in Chapter 4, we develop a decentralized adaptive controller that estimates the reference angular velocity and recovers the convergence results of the nonadaptive design. Although the presentation of the designs is based on the unit quaternion, we show that the same scheme is applicable to other parameterizations of $SO(3)$.

5.2 Kinematic Equation of Attitude Error

We consider a group of rigid bodies $i = 1, \cdots, N$, whose attitudes are represented by the 3×3 orthonormal attitude matrix $R_i \in SO(3)$[1]. The information flows of relative attitude measurements are represented by the undirected connected graph G. Similarly to Chapter 2, we assign an orientation to the graph to simplify the analysis.

If the ith and jth agents are connected by link k, we define the relative attitude of link k as

$$\tilde{R}^k := R_i^T R_j, \text{ if } k \in \mathcal{L}_i^+ \text{ and } k \in \mathcal{L}_j^- \tag{5.1}$$

and the relative angular velocity of link k as

$$\tilde{\omega}^k := {}^i\omega_j, \text{ if } k \in \mathcal{L}_i^+ \text{ and } k \in \mathcal{L}_j^- \tag{5.2}$$

where $\mathcal{L}_i^+$ ($\mathcal{L}_i^-$) denotes the set of links for which agent i is the positive (negative) end. This means that the relative angular velocity $\tilde{\omega}^k$ is represented in the frame of the positive end of link k. To simplify the notation, we introduce the $3N \times 3M$ *Rotational Incidence Matrix* $\bar{D}$, which consists of the 3×3 sub-blocks:

$$\bar{d}_{ik} := \begin{cases} -I_3 & \text{if } k \in \mathcal{L}_i^+ \\ (\tilde{R}^k)^T & \text{if } k \in \mathcal{L}_i^- \\ \mathbf{0}_3 & \text{otherwise,} \end{cases} \tag{5.3}$$

and note from (5.1) and (5.2) that

$$\tilde{\omega} = \bar{D}^T \omega^B \tag{5.4}$$

in which $\tilde{\omega} = [(\tilde{\omega}^1)^T, \cdots, (\tilde{\omega}^\ell)^T]^T$ and $\omega^B = [({}^1\omega_1)^T, \cdots, ({}^N\omega_N)^T]^T$.

The evolution of $\tilde{R}^k$ is obtained via direct differentiation with respect to the inertial frame [146]:

$$\frac{d\tilde{R}^k}{dt} = \widehat{\tilde{\omega}^k} \tilde{R}^k. \tag{5.5}$$

Let $q^k = \begin{bmatrix} q_0^k \\ q_v^k \end{bmatrix}$, $k = 1, \cdots, \ell$, be the unit quaternion representation parameterizing $\tilde{R}^k$, where q_0^k and q_v^k are the scalar and the vector parts, respectively. Then $\tilde{R}^k$ is related to q^k through the Rodriguez formula:

$$\tilde{R}^k = I_3 + 2(\widehat{q_v^k})^2 + 2q_0^k \widehat{q_v^k}, \tag{5.6}$$

and the kinematic equations of q^k are obtained from (5.5) and (5.6) as:

---

[1] For a brief introduction of attitude matrix and its representations, we refer to Appendix B.12.

$$J_k : \begin{cases} \dot{q}_0^k = -\frac{1}{2}(\tilde{\omega}^k)^T q_v^k \\ \dot{q}_v^k = \frac{1}{2} q_0^k \tilde{\omega}^k + \frac{1}{2} \widehat{\tilde{\omega}^k q_v^k}. \end{cases} \tag{5.7}$$

The following passivity property of the kinematics of unit quaternion, established in [84], is essential in the proof of Theorem 5.1 in Section 5.3.

Lemma 5.1 (Passivity of unit quaternion).
The unit quaternion kinematics (5.7) is passive *from $\tilde{\omega}^k$ to q_v^k with the storage function $V = (q_0^k - 1)^2 + |q_v^k|^2$.* □

We note that if $\tilde{R}^k$ equals I_3, which means that the agents connected by link k have the same attitude, then $q_v^k = [0,\ 0,\ 0]^T$ and $q_0^k = \pm\sqrt{1 - \|q_v^k\|^2} = \pm 1$. It follows that $q_0^k = \pm 1$ correspond to the same attitude in $SO(3)$. Without loss of generality, we choose to stabilize the $q_0^k = +1$ in the rest of this chapter. Furthermore, we assume that initially all the agents choose the quaternions with positive scalar parts.

When the columns of D are linearly dependent; that is, when the graph contains cycles, the unit quaternions q^k are also dependent. To see this, suppose that the graph contains $i = 1, \cdots, c$ cycles, each consisting of C_i links. For each cycle i, label the consecutive links by $k_i^1, \cdots, k_i^{C_i}$ and assign the directions of the links such that the positive end of each link is the negative end of the next link in the sequence (see Fig. 2.11). Then, from the definition of relative attitude in (5.1), it follows that

$$\prod_{j=1}^{C_i} \tilde{R}^{k_i^j} = I_3, \quad i = 1, \cdots, c \tag{5.8}$$

or in terms of quaternions,

$$q^{k_i^1} \circ q^{k_i^2} \circ \cdots \circ q^{k_i^{C_i}} = (\pm 1\ 0\ 0\ 0)^T, \quad i = 1, \cdots, c \tag{5.9}$$

where q^k is the quaternion parametrization of $\tilde{R}^k$ of link k and $\circ$ denotes the quaternion multiplication[2].

5.3 Passivity-based Group Attitude Agreement

The dynamics of the attitude of the ith agent is given by the Euler equation:

$${}^i\mathcal{J}_i\, {}^i\dot{\omega}_i + {}^i\omega_i \times {}^i\mathcal{J}_i\, {}^i\omega_i = {}^i\tau_i, \tag{5.10}$$

where ${}^i\mathcal{J}_i$ is the inertia matrix and ${}^i\tau_i$ is the control input in the ith frame. The objective is to develop a decentralized coordination law for each agent that depends

[2] Given two unit quaternions $p^T = [p_0, p_v^T]$ and $q^T = [q_0, q_v^T]$, where p_0, q_0 are scalars and p_v, q_v are 3×1 vectors, the quaternion product is defined as $p \circ q = \begin{pmatrix} q_0 p_0 - q_v^T p_v \\ \widehat{p_v} q_v + p_0 q_v + q_0 p_v \end{pmatrix}$.

only on its angular velocity and on its relative attitude with respect to its neighbors to guarantee the following behaviors:

B1) Each agent achieves the same attitude as its neighbors in the limit; that is $\lim_{t\to\infty} R_i^T R_j = I_3$, $i, j = 1, \cdots, N$.

B2) The angular velocity of each agent converges to a reference angular velocity $\omega^d(t)$*; that is* $\lim_{t\to\infty} |{}^i\omega_i - \omega^d(t)| = 0$, $i = 1, \cdots N$.

We assume that $\omega^d(t)$ and $\dot{\omega}^d(t)$ are continuous and uniformly bounded. In objective B2, the agents follow $\omega^d(t)$ in their individual body frames. When objective B1 is achieved, B2 means that the agents rotate synchronously.

To achieve objectives B1 and B2, we follow the design procedure in Section 2.3. We first design an internal feedback loop ${}^i\tau_i$ for each agent $i = 1, \cdots, N$ that renders its dynamics passive from an external input signal u_i left to be designed, to the angular velocity error

$$\Delta\omega_i := {}^i\omega_i - \omega^d(t). \tag{5.11}$$

Indeed, the controller

$${}^i\tau_i = {}^i\mathscr{J}_i\,\dot{\omega}^d + \omega^d \times {}^i\mathscr{J}_i\,{}^i\omega_i - f_i\Delta\omega_i + u_i, \tag{5.12}$$

where f_i is a positive constant, yields the error dynamics system:

$$\mathscr{H}_i: \quad {}^i\mathscr{J}_i\Delta\dot{\omega}_i + \Delta\omega_i \times {}^i\mathscr{J}_i\,{}^i\omega_i = -f_i\Delta\omega_i + u_i, \tag{5.13}$$

which is strictly passive from u_i to $\Delta\omega_i$ with the storage function $V_i(\Delta\omega_i) = \frac{1}{2}(\Delta\omega_i)^T\,{}^i\mathscr{J}_i\Delta\omega_i$. The stability analysis below relies on this passivity property, and not on the particular form of the control law (5.12). To achieve further design flexibility, other controllers that achieve passivity from u_i to $\Delta\omega_i$ may be employed.

Next, we design the external feedback u_i of the form:

$$u_i = \sum_{l\in\mathscr{L}_i^+} q_v^l - \sum_{p\in\mathscr{L}_i^-} q_v^p. \tag{5.14}$$

To synthesize this external feedback u_i, each agent i obtains its neighbors' relative attitudes with respect to its own frame, parameterizes them by unit quaternions, and then adds up the vector parts of the unit quaternions. Thus, the control law (5.12) with the external input (5.14) is implementable in a decentralized fashion.

To analyze the stability of the closed-loop system, we denote by J_k, $k = 1, \cdots, \ell$, the kinematic equation (5.7) for the kth link with input $\tilde{\omega}^k$ and output q_v^k, and let

$$\mathscr{F} = \text{diag}\{f_1, \cdots, f_N\} \otimes I_3 \qquad \mathscr{J} = \text{diag}\{{}^1\mathscr{J}_1, \cdots, {}^N\mathscr{J}_N\} \tag{5.15}$$

$$\mathscr{E}_\omega = [\Delta\omega_1^T, \cdots, \Delta\omega_N^T]^T \quad u = [u_1^T, \cdots, u_N^T]^T \quad \tau^B = [{}^1\tau_1^T, \cdots, {}^N\tau_N^T]^T \tag{5.16}$$

$$q_0 = [q_0^1, \cdots, q_0^\ell]^T \quad q_v = [(q_v^1)^T, \cdots, (q_v^\ell)^T]^T. \tag{5.17}$$

Using this notation, and rewriting (5.14) as

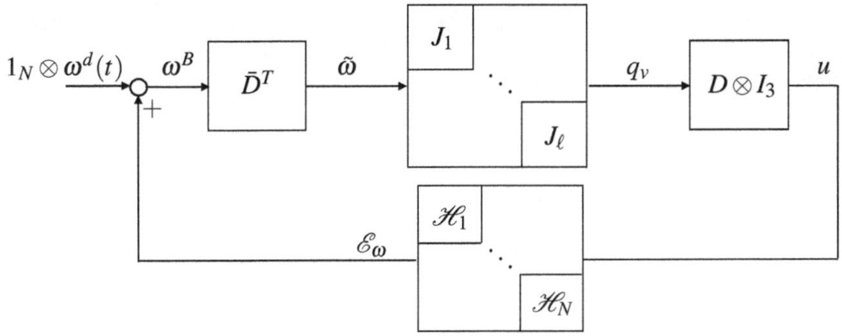

Fig. 5.1 Interconnection of the quaternion kinematics (5.7) and the error dynamics (5.13). J_k denotes the quaternion kinematics (5.7) and $\mathscr{H}_i$ is as (5.13).

$$u = (D \otimes I_3) q_v \tag{5.18}$$

where D is as in (1.21), we obtain the closed-loop block diagram in Fig. 5.1.

Theorem 5.1 below makes use of the passivity properties of the feedforward and the feedback paths in Fig. 5.1, and analyzes stability properties of the origin in the $\{\mathscr{E}_\omega,\ q_v\}$-space. Before proceeding with this analysis, however, we point to an obstacle to the *global* convergence to the origin which arises when the graph contains cycles. To see this note that the set of equilibria for the closed-loop system in Fig. 5.1 is given by:

$$\mathscr{O} = \{(\mathscr{E}_\omega, q_0, q_v) |\ \mathscr{E}_\omega = 0,\ \text{constraint (5.9) and } (D \otimes I_3) q_v = 0\}, \tag{5.19}$$

where $\mathscr{E}_\omega = 0$ implies ${}^i\omega_i = \omega^d$, which means that each agent achieves the reference angular velocity. If the graph is acyclic, that is, if the columns of D are linearly independent, then the third constraint $(D \otimes I_3) q_v = 0$ implies $q_v = 0$, which is the desired relative attitude where all agents are aligned. However, when the graph contains cycles, the null space of $(D \otimes I_3)$ is nontrivial, and the set in (5.19) contains additional equilibria. In Theorem 5.1 below we prove that all trajectories converge to the equilibrium set $\mathscr{O}$.

Theorem 5.1. *For the agents $i = 1, \cdots, N$ defined by the dynamics (5.10), the control laws (5.12) and (5.18) guarantee that the signals $\{\mathscr{E}_\omega,\ q_0,\ q_v\}$ converge to the set of equilibria in (5.19) and the equilibrium $\{\mathscr{E}_\omega,\ q_0,\ q_v\} = \{0, +1_\ell, 0\}$ is stable. If D is full column rank, that is, if the graph is acyclic, then $q_v \to 0$ and $\mathscr{E}_\omega \to 0$ as $t \to \infty$.* □

Proof. We first make use of Lemma 5.1 and prove that the feedforward path in Fig. 5.1 is passive from ω^B to $-u$. To show this, we take the storage function

$$V_u = (q_0 - 1_\ell)^T (q_0 - 1_\ell) + q_v^T q_v, \tag{5.20}$$

and note that its derivative yields

$$\dot{V}_u = \tilde{\omega}^T q_v. \tag{5.21}$$

We further rewrite (5.21) from (5.4) as

$$\dot{V}_u = (\omega^B)^T (\bar{D} q_v). \tag{5.22}$$

It then follows from Lemma 5.2 below that

$$\dot{V}_u = (\omega^B)^T (-(D \otimes I_3) q_v) = (\omega^B)^T (-u), \tag{5.23}$$

which proves passivity from ω^B to $-u$.

Having proven passivity of the feedforward path using the storage function V_u, we now take the Lyapunov function

$$V = V_u + \frac{1}{2} \mathscr{E}_\omega^T \mathscr{J} \mathscr{E}_\omega, \tag{5.24}$$

and note that its derivative along (5.7), (5.10), (5.12) and (5.18), satisfies

$$\dot{V} = -(\omega^B)^T u + \mathscr{E}_\omega^T u - \mathscr{E}_\omega^T \mathscr{F} \mathscr{E}_\omega \tag{5.25}$$

$$= -(1_N \otimes \omega^d)^T u - \mathscr{E}_\omega^T \mathscr{F} \mathscr{E}_\omega. \tag{5.26}$$

Recalling that $u = (D \otimes I_3) q_v$ from (5.18) and that $(D \otimes I_3)^T (1_N \otimes \omega^d) = 0$, we obtain from (5.26) that

$$\dot{V} = -\mathscr{E}_\omega^T \mathscr{F} \mathscr{E}_\omega \leq 0, \tag{5.27}$$

which proves that the equilibrium $\{\mathscr{E}_\omega,\ q_0,\ q_v\} = \{0, +1_\ell, 0\}$ is stable. It then follows that all the signals $(\mathscr{E}_\omega, u, q_0, q_v)$ are bounded. It then follows from Theorem B.5 that $\mathscr{E}_\omega \to 0$, that is $|{}^i\omega_i - \omega^d(t)| \to 0$. Next, we show that $u \to 0$. To this end, we note that $\Delta\ddot{\omega}_i$ is continuous and uniformly bounded from (5.13) because the signals $\{\dot{u}_i,\ \Delta\dot{\omega}_i,\ {}^i\mathscr{J}_i,\ \omega^d,\ \dot{\omega}^d\}$ are bounded and continuous. Since $\Delta\omega_i \to 0$ and $\Delta\ddot{\omega}_i$ is continuous and bounded, it follows from Theorem B.4 that $\Delta\dot{\omega}_i \to 0$, which implies from (5.13) that $u_i \to 0$.

Finally, we note from (5.18) that $u \to 0$ means that q_v converges to the null space $\mathscr{N}(D \otimes I_3)$. In particular, if D is full column rank, which means there are no cycles in the graph, then $\mathscr{N}(D \otimes I_3) = 0$ and, thus, $q_v \to 0$. □

Lemma 5.2 below was used in the proof for Theorem 5.1 to establish the passivity of the feedforward path in Fig. 5.1.

Lemma 5.2 (Rotation invariance).
The vector part q_v of the unit quaternion satisfies:

$$(D \otimes I_3) q_v = -\bar{D} q_v \tag{5.28}$$

where D is as in (1.21) and $\bar{D}$ is as in (5.3). □

Proof. Define $\bar{D}_i = [\bar{d}_{i1}, \cdots, \bar{d}_{i\ell}]$, where $\bar{d}_{ik}$ is as in (5.3) and note that

$$
\begin{aligned}
-\bar{D}_i q_v &= -\sum_{k=1}^{\ell} \bar{d}_{ik} q_v^k \\
&= \sum_{l \in \mathscr{L}_i^+} q_v^l + \sum_{p \in \mathscr{L}_i^-} (\tilde{R}^p)^T(-q_v^p).
\end{aligned} \tag{5.29}
$$

Substituting (5.6) into (5.29) and using (1.21), we further obtain

$$
\begin{aligned}
-\bar{D}_i q_v &= \sum_{l \in \mathscr{L}_i^+} q_v^l - \sum_{p \in \mathscr{L}_i^-} q_v^p \\
&= (D_{i,\cdot} \otimes I_3) q_v,
\end{aligned} \tag{5.30}
$$

where $D_{i,\cdot}$ represents the ith row of D, and note that (5.30) is equivalent to (5.28).

□

5.4 Other Representations of $SO(3)$

In this section, we show that the unit quaternion can be replaced by any other representation of $SO(3)$, if this representation satisfies the passivity property (Lemma 5.1) and the rotation invariance property (Lemma 5.2). Among the three-parameter representations that satisfy these two properties are the vector quaternion, Gibb's vector, modified Gibb's vector and the unit equivalent axis/angle.

To present a unified storage function for these representations, we let h^k be a unit vector along the equivalent axis of $\tilde{R}^k$ and let $\tilde{\theta}^k$ be the equivalent angle. We then write the parameter vector s^k in the form:

$$
s^k = \gamma(\tilde{\theta}^k) h^k, \tag{5.31}
$$

where $\gamma(\cdot)$ is a first-third quadrant odd nonlinearity and its specific form depends on the representation as follows:

$$
\gamma_{vq}(\tilde{\theta}^k) = \sin(\frac{\tilde{\theta}^k}{2}) \quad \text{(vector quaternion)} \tag{5.32}
$$

$$
\gamma_{G}(\tilde{\theta}^k) = \tan(\frac{\tilde{\theta}^k}{2}) \quad \text{(Gibb's vector)} \tag{5.33}
$$

$$
\gamma_{mG}(\tilde{\theta}^k) = \tan(\frac{\tilde{\theta}^k}{4}) \quad \text{(modified Gibb's vector)} \tag{5.34}
$$

$$
\gamma_{e}(\tilde{\theta}^k) = \tilde{\theta}^k \quad \text{(equivalent axis/angle).} \tag{5.35}
$$

The following lemma unifies the earlier passivity results in [84] and [139]:

Lemma 5.3. *The kinematics of three-parameter parametrization s^k in (5.31), where $\gamma(\cdot)$ is an odd first-third quadrant nonlinearity, is passive from $\tilde{\omega}^k$ to s^k.* □

Proof. The kinematics of $\tilde{\theta}^k$ and h^k are given by [146]:

$$\dot{\tilde{\theta}}^k = (\tilde{\omega}^k)^T h^k \tag{5.36}$$

$$\dot{h}^k = -\frac{1}{2}(\widehat{h^k} + \cot\frac{\tilde{\theta}^k}{2}(\widehat{h^k})^2)\tilde{\omega}^k. \tag{5.37}$$

A valid storage function for all the representations defined in (5.31) is chosen as

$$V_p = \int_0^{\tilde{\theta}^k} \gamma(\theta)d\theta. \tag{5.38}$$

Since $\gamma(\cdot)$ is a first-third quadrant odd function, V_p is an even and positive definite function. The derivative of V_p is

$$\begin{aligned}\dot{V}_p &= \gamma(\tilde{\theta}^k)\dot{\tilde{\theta}}^k \\ &= (s^k)^T \tilde{\omega}^k\end{aligned} \tag{5.39}$$

which proves the passivity from $\tilde{\omega}^k$ to s^k. □

We now show that s^k satisfies "rotation invariance", that is,

$$(D \otimes I_3)s = -\bar{D}s. \tag{5.40}$$

where $s = [(s^1)^T, \cdots, (s^\ell)^T]^T$.

Lemma 5.4. *The three-parameter parametrization s^k satisfies the property of rotation invariance (5.40).* □

Proof. The property of rotation invariance originates from the fact that the equivalent axis of an attitude matrix is an eigenvector of that matrix associated with the eigenvalue 1, that is $\tilde{R}^k h^k = h^k$. To prove the identity (5.40), we follow a procedure similar to that in (5.29)-(5.30), and obtain

$$-\bar{D}_i s = (D_{i,\cdot} \otimes I_3)s, \tag{5.41}$$

which is equivalent to (5.40). □

Since s^k satisfies the properties of passivity and rotation invariance, the framework in Theorem 5.1 follows directly. Indeed, the internal feedback law (5.12) remains the same while the external control law u_i is now defined as

$$u_i = \sum_{l \in \mathscr{L}_i^+} s^l - \sum_{p \in \mathscr{L}_i^-} s^p, \tag{5.42}$$

which can be synthesized similarly as for the unit quaternion.

However, because s^k is a three-parameter representation and is only locally one-to-one and onto mappings of attitude matrix, it is not a globally nonsingular three-parameter representation of $SO(3)$. In other words, the Jacobian that maps differential changes in s^k to differential changes in the attitude matrix is singular for some orientations. For example, the vector quaternion ($h^k \sin(\frac{\tilde{\theta}^k}{2})$) and the Gibb's vector ($h^k \tan(\frac{\tilde{\theta}^k}{2})$) are singular when $\tilde{\theta}^k = \pi$. This singularity is not a physical one in the sense that relative attitude can not move in some direction. It is a mathematical deficiency due to the local nature of three-parameter representation. In that singular situation, s^k is not a valid parametrization and other representation that is nonsingular at that point might be used.

5.5 Attitude Coordination in the Plane

We now reduce the controller from the three dimensional space to the plane. This means that the first two entries of ${}^i\omega_i$ are zero and, thus, (5.10) is

$$\mathscr{J}_i {}^i\dot{\omega}_i = {}^i\tau_i \tag{5.43}$$

where $\mathscr{J}_i \in \mathbb{R}$ is the inertia of agent i in the plane, ${}^i\omega_i \in \mathbb{R}$ is the angular velocity and ${}^i\tau_i \in \mathbb{R}$ is the control torque of agent i. Likewise, the $\bar{D}$ matrix in (5.3) now becomes $-D$. We note that for each agent the rotation axis is the z-axis, which implies that $q_0^k = \cos\frac{\tilde{\theta}^k}{2}$ and $q_v^k = [0\ 0\ \sin\frac{\tilde{\theta}^k}{2}]^T$, where $\tilde{\theta}^k \in [-\pi, \pi)$ is the relative attitude of link k connecting the ith and jth agents. We thus obtain from (5.14) the control law

$$u_i = \sum_{l \in \mathscr{L}_i^+} \sin(\frac{\tilde{\theta}^l}{2}) - \sum_{p \in \mathscr{L}_i^-} \sin(\frac{\tilde{\theta}^p}{2}). \tag{5.44}$$

To see the equilibria that result from this controller, consider the special case of a ring interconnection in Fig. 2.11. In this case, the incidence matrix D has the cyclic structure

$$D = \begin{pmatrix} -1 & 0 & 0 & \cdots & 0 & 1 \\ 1 & -1 & 0 & \cdots & 0 & 0 \\ 0 & 1 & -1 & \cdots & 0 & 0 \\ \vdots & \vdots & \ddots & \ddots & \vdots & \vdots \\ 0 & 0 & \cdots & 1 & -1 & 0 \\ 0 & 0 & 0 & \cdots & 1 & -1 \end{pmatrix} \tag{5.45}$$

whose null space is spanned by 1_N. Then from (5.9) and by letting u_i in (5.14) to be zero, we obtain the equilibrium conditions

$$\sin\frac{\tilde{\theta}^i}{2} = \sin\frac{\tilde{\theta}^j}{2} \quad i, j = 1, \cdots, N \tag{5.46}$$

and

$$\sum_{i=1}^{N} \tilde{\theta}^i = c \cdot 2\pi \quad c = 0, \cdots, N-1. \tag{5.47}$$

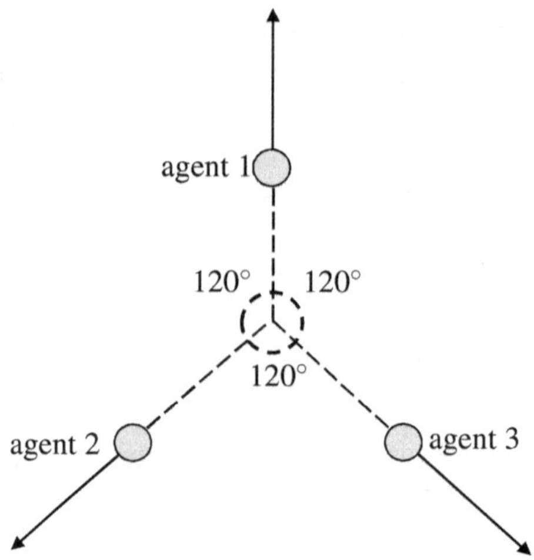

Fig. 5.2 One of the equilibria configurations for a ring connection of three agents. Arrows denote the headings of the agents.

In addition to the desired equilibrium where $\tilde{\theta}^i = 0$, $i = 1, \cdots, N$, (5.46)-(5.47) possess other solutions, including

$$\tilde{\theta}^i = \frac{2\pi}{N} \quad i = 1, \cdots, N \tag{5.48}$$

which is depicted in Fig. 5.2 for the case of $N = 3$ agents.

For $N = 3$, however, a study of the Jacobian linearization shows that the undesired equilibria are unstable. This is similar to the formation control example in Example 2.4, where the undesired collinear equilibria are unstable. We thus conclude generic convergence to the desired equilibrium from all initial conditions except for those that lie on the stable manifolds of the unstable equilibria.

For other representations $s^k = \gamma(\theta^k)h^k$ defined in Section 5.4, the controller u_i in (5.44) takes the form

$$u_i = D_i[\gamma(\tilde{\theta}^1), \cdots, \gamma(\tilde{\theta}^M)]^T. \tag{5.49}$$

and the equilibria are now described by

$$\gamma(\tilde{\theta}^i) = \gamma(\tilde{\theta}^j), \quad i, j = 1, \cdots, N \tag{5.50}$$

and (5.47). Further investigation shows that (5.47) and (5.50) always possess the equilibrium (5.48) as well as the desired one.

5.6 Adaptive Design for Reference Angular Velocity Recovery

The design in Section 5.3 assumes that the reference angular velocity $\omega^d(t)$ is available to each agent. We now consider the situation where only the leader, say, the first agent $i = 1$, possesses the $\omega^d(t)$ information. Since the structure in Fig. 5.1 is similar to that in Fig. 2.2, we follow procedures similar to those in Chapters 3 and 4 to develop an adaptive design with which the remaining agents estimate $\omega^d(t)$. For conciseness, we only consider the approach in Chapter 4. We then assume that $\omega^d(t) \in \mathbb{R}^3$ is parameterized as

$$\omega^d(t) = \sum_{j=1}^{r} \phi^j(t)\beta^j \tag{5.51}$$

where $\phi^j(t)$ are scalar basis functions available to each agent and $\beta^j \in \mathbb{R}^3$ are column vectors available only to the leader $i = 1$.

The other agents estimate the unknown β^j by $\bar{\beta}^j$, and construct ${}^i\bar{\omega}_i^d(t)$ from

$${}^i\bar{\omega}_i^d(t) = \sum_{j=1}^{r} \phi^j(t)\bar{\beta}_i^j = (\Phi(t)^T \otimes I_3)\bar{\beta}_i \quad i = 2,\cdots,N \tag{5.52}$$

where $\Phi(t) := [\phi^1(t),\cdots,\phi^r(t)]^T$ and $\bar{\beta}_i := [(\bar{\beta}_i^1)^T,\cdots,(\bar{\beta}_i^r)^T]^T$.

The adaptive design employs the feedback law (5.14) in Section 5.3, and modifies (5.12) as

$${}^1\tau_1 = {}^1\mathscr{J}_1\, \dot{\omega}_d + \omega_d \times {}^1\mathscr{J}_1\, {}^1\omega_1 - f_1 \Delta\omega_i + u_1 \tag{5.53}$$

$${}^i\tau_i = {}^i\mathscr{J}_i\, {}^i\dot{\bar{\omega}}_i^d + {}^i\bar{\omega}_i^d \times {}^i\mathscr{J}_i\, {}^i\omega_i - f_i \Delta\bar{\omega}_i + u_i, \quad i = 2,\cdots,N \tag{5.54}$$

where ${}^i\bar{\omega}_i^d$ is now obtained from (5.52) and

$$\Delta\bar{\omega}_i := {}^i\omega_i - {}^i\bar{\omega}_i^d.$$

The update law for the parameter $\bar{\beta}_i$ is

$$\dot{\bar{\beta}}_i = \Lambda_i(\Phi(t) \otimes I_3)u_i \tag{5.55}$$

in which $\Lambda_i = \Lambda_i^T > 0$ is the adaptive gain matrix and u_i is as in (5.14). With this update law, we recover the convergence result proven for the nonadaptive design in Section 5.3.

Theorem 5.2. *Consider the agents $i = 1,\cdots,N$ defined by the dynamics (5.10) and suppose the reference angular velocity $\omega^d(t)$ is available only to agent 1. The control laws (5.53), (5.54) and (5.18) together with the update law (5.55) guarantee that the signals $\{q_0, q_v, \Delta\omega_1, \Delta\bar{\omega}_i, \bar{\beta}_i\}$, $i = 2,\cdots,N$ are bounded. Furthermore, $\Delta\omega_1$ and $\Delta\bar{\omega}_i$, $i = 2,\cdots,N$ converge to zero and q_v converges to the equilibria where*

$(D \otimes I_3)q_v = 0$ subject to the constraint in (5.9). In particular, if $q_v \to 0$, then $|{}^i\omega_i - \omega^d| \to 0$. □

The proof for this theorem is given in Appendix A.4. Theorem 5.2 proves that when the attitudes of the rigid bodies reach agreement, that is, $q_v \to 0$, ${}^i\omega_i$ converges to the reference angular velocity $\omega^d(t)$. Such a case occurs when D is full column rank, which means that the graph is acyclic. If $\Phi(t)$ satisfies the same PE condition as in (4.23), we can further show the parameter convergence $\bar{\beta}_i \to \beta$ from Lemma 4.1.

5.7 Simulation Results

In this section, we present an example of three agents, where each of them is a neighbor of the other two. We take the moment of inertia in the body frame of each agent to be

$$ {}^i\mathscr{J}_i = \begin{bmatrix} 1 & 0 & 0 \\ 0 & 0.63 & 0 \\ 0 & 0 & 0.85 \end{bmatrix}, \quad i = 1,2,3. \tag{5.56}$$

The first simulation shows the convergence to the desired equilibrium and to the reference angular velocity. The second simulation illustrates the adaptive scheme.

5.7.1 Nonadaptive Design

In this simulation, we show that three agents converge to the same attitude and the same reference angular velocity by only relative attitude information. The relative orientations $\tilde{R}^1(0)$, $\tilde{R}^2(0)$ and $\tilde{R}^3(0)$ are parameterized by the initial conditions:

$$\begin{aligned} q^1(0) &= [0.9289\ 0.0559\ 0.3652\ \ -0.0260]^T \\ q^2(0) &= [0.8699\ 0.0165\ 0.4917\ 0.0353]^T \\ q^3(0) &= [0.6285\ \ -0.0896\ \ -0.7720\ \ -0.0316]^T. \end{aligned}$$

The desired angular velocity ω^d is chosen as $[0.0429\ \ 0.0059\ \ 0.1413]^T$. The initial angular velocity in each agent's frame is zero. Fig. 5.3 shows that the desired relative attitude is indeed achieved with the design (5.12) and (5.14) since the scalar parts of q^k, $k = 1,2,3$, converge to 1. Moreover, $|{}^i\omega_i - \omega^d|$, $i = 1,2,3$, converge to zero, which in turn implies that ${}^i\omega_i \to \omega^d$.

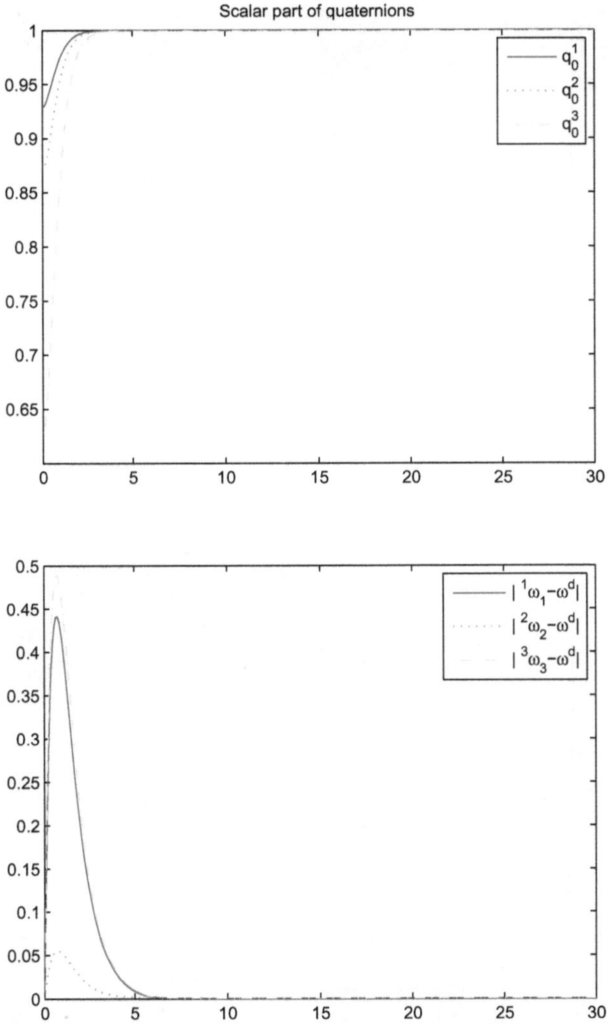

Fig. 5.3 Desired relative attitude convergence and reference angular velocity convergence.

5.7.2 Adaptive Design

In this simulation, we show the convergence of the adaptive design with the same constant desired angular velocity ω^d as in the nonadaptive design. By applying the update law in (5.55) and control laws in (5.14), (5.53) and (5.54), we recover the convergence results in the nonadaptive design. Fig. 5.4 shows convergence to the reference angular velocity and convergence to the desired relative attitude where all the agents are aligned.

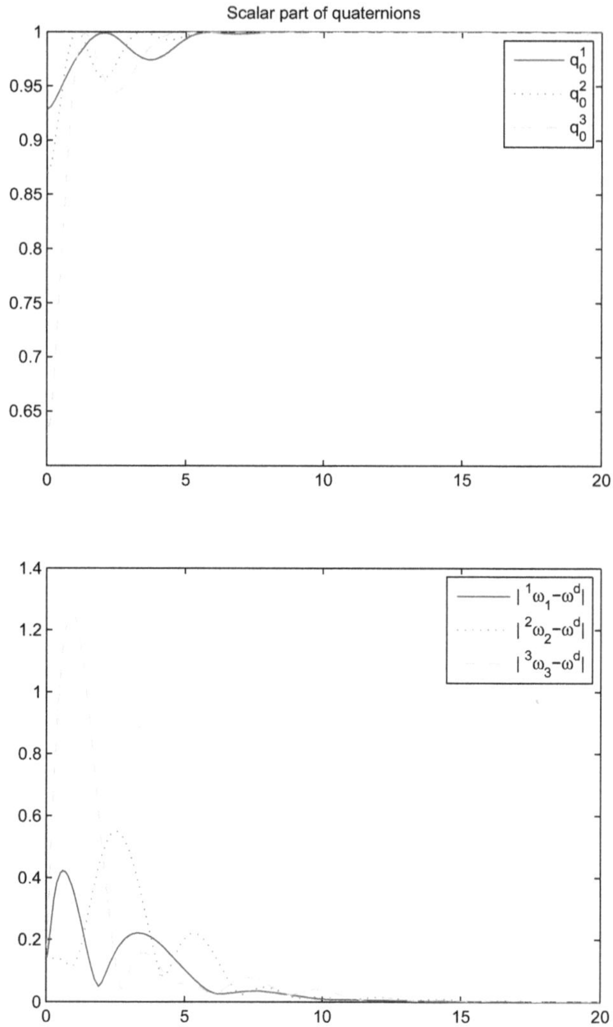

Fig. 5.4 Desired relative attitude convergence and reference angular velocity convergence in the adaptive design.

5.8 Summary

We studied an attitude alignment problem where the inertial frame information is not available to the agents. The decentralized controllers depend on the relative orientations and local angular velocities, and guarantee boundedness of the trajectories as well as their convergence to the set of equilibria. We further considered the situation where the reference angular velocity information is available only to one agent, and developed an adaptive design with which the other agents recover this information.

5.9 Related Literature

The attitude coordination problem has been considered in a number of publications. For example, reference [66] discussed several coordinated attitude control strategies, such as a leader-follower strategy and a virtual desired attitude strategy, and illustrated by simulations the advantage of coordination in the presence of disturbances. Reference [141] introduced a decentralized controller to achieve the convergence of each agent to the same commanded desired attitude. In [75], the authors introduced an attitude coordination algorithm applicable to a ring communication graph. The results in [75] were extended into a more general graph in [108]. Both [75] and [108], however, rely on the inertial frame information. Reference [38] used the Modified Rodriguez Parameter to represent the absolute attitudes of rigid bodies. In [20], the authors presented a leader/follower strategy of attitude synchronization by employing absolute attitudes and angular velocity observers. Unlike the controllers in [141, 75, 108, 38, 66], which require the knowledge of the inertial frame, [95] presented a design that only depends on the relative orientation among neighboring agents. However, the design of [95] requires that the communication topology of the agents form a string connection. Reference [115] considered using relative angular velocity in attitude synchronization. The work in [58] considered kinematic control of attitude synchronization. Since the agent kinematics are relative degree one, the attitude synchronization can be achieved with strongly connected graphs. Recent research in [31] represented the attitude dynamics in the Lagrangian form and used contraction analysis to study coordinated attitude tracking with directed graphs.

Chapter 6
The Agreement of Euler-Lagrange Systems

6.1 Introduction

In this chapter, we further extend the passivity approach from point robots to agents modeled as Euler-Lagrange systems. The Euler-Lagrange model is prevalently used to model multi-link robots and encompasses the double integrator model in (2.16) as a special case. We first apply the design techniques in Chapter 2 and present a nominal design that achieves the agreement of the agents. We next consider the case where the dynamics of each agent are subject to a class of parameterized uncertainty. Such uncertainty may come from unknown parameters in agent dynamics or external disturbances, such as wind and friction. We develop adaptive control laws that compensate for the uncertainty and achieve objectives A1 and A2. The closed-loop structure considered in this chapter is shown in Fig. 6.1, where the blocks of the network information flow and the agent dynamics are the feedforward and feedback subsystems in Fig. 2.2, respectively.

To develop adaptive control laws, we draw on earlier results in single robot trajectory tracking and present two approaches. The first approach is motivated by [138, 143, 68], where a small *cross term* was added to a Lyapunov function to achieve adaptive trajectory tracking for *single* robot. Applying a similar approach to the agreement of *multiple* Euler-Lagrange systems, we obtain a Lyapunov function with a cross term of the storage functions in (2.35). Different from the Lyapunov function considered in Section 2.4, this Lyapunov function yields a negative definite derivative, which allows us to develop an adaptive design that recovers objectives A1 and A2 in the presence of uncertainty. The second approach explores the design flexibility offered by the passivity framework. We modify the nominal design by feeding forward the external feedback u to the velocity error y, which ensures that the Lyapunov function in Section 2.4 yields a negative definite derivative. The resulting closed-loop system takes a form similar to the controller in [121], where a *virtual reference trajectory* was introduced for adaptive design of single robot trajectory tracking. In our problem, the virtual reference velocity for each agent is the aggregation of the differences between itself and its neighbors. The results in

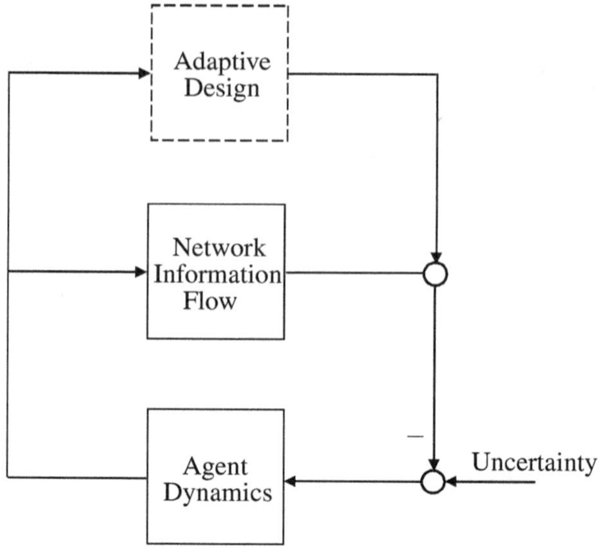

Fig. 6.1 The passivity structure considered in this chapter. Individual agent dynamics are now subject to uncertainty, such as wind disturbance and uncertain parameters in agent model. We will develop adaptive control laws that compensate for the uncertainty and achieve objectives A1 and A2.

this chapter extend the two widely-used approaches [138] and [121] in single robot trajectory tracking to the coordination of multiple robots.

6.2 The Nominal System

We consider a group of N agents, each modeled as an Euler-Lagrange system

$$M_i(x_i)\ddot{x}_i + C_i(x_i,\dot{x}_i)\dot{x}_i = \tau_i \quad i = 1,\cdots,N \tag{6.1}$$

where $x_i \in \mathbb{R}^p$. The goal is to achieve the agreement of x_i's and objective A1 in (2.1). We assume that the reference velocity $v(t)$ in objective A1 is bounded and let

$$\mu_v = \sup_{t\geq 0} |v(t)|. \tag{6.2}$$

We will consider the Euler-Lagrange systems that satisfy the following properties [70, 105]:

Property 6.1. $M_i(x_i)$ is positive definite and there exist two positive constants γ_{m_i} and γ_{M_i} such that

$$\gamma_{m_i} I_p < M_i(x_i) < \gamma_{M_i} I_p. \tag{6.3}$$

□

Property 6.2. The matrix $C_i(x_i,\dot{x}_i)$ is bounded in x_i and linear in $\dot{x}_i$ and thus satisfies

$$||C_i(x_i,\dot{x}_i)|| \leq f_{ci}|\dot{x}_i| \tag{6.4}$$

where $f_{ci} > 0$. □

To achieve the control objective, we follow the design steps in Section 2.3. We first introduce an internal feedback

$$\tau_i = -K_i(\dot{x}_i - v(t)) + M_i(x_i)\dot{v}(t) + C_i(\dot{x}_i, x_i)v(t) + u_i, \quad K_i = K_i^T > 0 \tag{6.5}$$

which, together with a change of variable $\xi_i = \dot{x}_i - v(t)$, renders the following nominal system

$$\dot{x}_i = y_i + v(t) \tag{6.6}$$

$$\mathscr{H}_i : \begin{cases} M_i(x_i)\dot{\xi}_i = -C_i(x_i,\dot{x}_i)\xi_i - K_i\xi_i + u_i \\ y_i = \xi_i. \end{cases} \tag{6.7}$$

Choosing

$$S_i(\xi_i) = \frac{1}{2}\xi_i^T M_i(x_i)\xi_i \tag{6.8}$$

as a storage function and using the skew symmetry property in (1.35), we verify that $\mathscr{H}_i$ is strictly passive from u_i to y_i. Thus, $\mathscr{H}_i$ is of the form (2.11) with the required passivity property and satisfies (2.12) and (2.13).

We next design u_i. We define z_k as in (2.2). Following the result in Section 2.6.1, we choose u_i as in (2.68) with ψ_k defined in (2.72), which yields

$$u = -(L_\Delta \otimes I_p)x = -(D\Delta \otimes I_p)z \tag{6.9}$$

where Δ is defined in (2.73). Recall that for the agreement problem, $u \to 0$ is equivalent to $z \to 0$, since Property 2.1 is satisfied with ψ_k in (2.72).

Although the internal feedback design in (6.5) and the choice of u in (6.9) follow techniques in Section 2.3, the stability results in Theorem 2.1 are not directly applicable here. This is because the $\mathscr{H}_i$ dynamics in (6.7) contain $x_i(t)$ signal, thereby making $\mathscr{H}_i$ a non-autonomous system of ξ_i and z_k. We next demonstrate uniformly globally asymptotic stability of the origin of (z,ξ) by applying the Nested Matrosov Theorem. The convergence to the origin of (z,ξ) means that the agreement of x_i's and reference velocity tracking are achieved.

Theorem 6.1. *Consider the agent dynamics in (6.1) with τ_i defined in (6.5) and u in (6.9). Then the origin of (ξ,z) is uniformly globally asymptotically stable.* □

Proof. We take the Lyapunov function

$$V_1 = \sum_{i=1}^{N} S_i(\xi_i) + \sum_{k=1}^{\ell} P_k(z_k) \tag{6.10}$$

where $S_i(\xi_i)$ and $P_k(z_k)$ are given in (6.8) and (2.71), respectively. Using (6.7), (2.6) and (6.9), we obtain the derivative of V_1 as

$$\dot{V}_1 = \sum_{i=1}^{N}(-\xi_i^T K_i \xi_i + \xi_i^T u_i) + \sum_{k=1}^{\ell} \delta_k z_k^T \dot{z}_k \tag{6.11}$$

$$= -\sum_{i=1}^{N} \xi_i^T K_i \xi_i - y^T(D\Delta \otimes I_p)z + z^T(\Delta \otimes I_p)(D^T \otimes I_p)y \tag{6.12}$$

$$= -\sum_{i=1}^{N} \xi_i^T K_i \xi_i := Y_1 \leq 0 \tag{6.13}$$

which implies the uniform stability of the origin of (ξ, z).

To establish uniform asymptotic stability, we define an auxiliary function

$$V_2 = -u^T \xi \tag{6.14}$$

and claim

$$Y_1 = 0 \Longrightarrow \dot{V}_2 := Y_2 \leq 0. \tag{6.15}$$

To see this claim, we note

$$Y_2 = -\dot{u}^T \xi - u^T \dot{\xi}. \tag{6.16}$$

When $\dot{V}_1 = 0$, ξ is zero, which implies that the first term in (6.16) vanishes. It follows from (6.7) that $\dot{\xi}_i = M_i^{-1}(x_i)u_i$ when $\xi = 0$, which means

$$Y_2\big|_{Y_1=0} = -\sum_{i=1}^{N} u_i M_i^{-1}(x_i) u_i \leq 0. \tag{6.17}$$

Using Property 6.1, we conclude (6.15). We further note that $Y_1 = Y_2 = 0$ implies $\xi = 0$ and $u = 0$. Recall that for agreement problem Property 2.1 is satisfied and thus $u = 0$ is equivalent to $z = 0$ (see (2.34)). Then all assumptions of the Nested Matrosov Theorem in Theorem B.5 are satisfied and we conclude uniformly globally asymptotic stability of the origin of (ξ, z). □

6.3 The Uncertain System

Suppose that the dynamics of each agent are now subject to a class of uncertainty $d_i(t, x_i, \dot{x}_i)$, parameterized by

$$d_i(t, x_i, \dot{x}_i) = Y_i(t, x_i, \dot{x}_i)\theta_i, \tag{6.18}$$

where $\theta_i \in \mathbb{R}^r$ is an unknown constant vector and $Y_i(t, x_i, \dot{x}_i) \in \mathbb{R}^{p \times r}$ is a regressor matrix available to agent i. Then the agent's model becomes

$$M_i(x_i)\ddot{x}_i + C_i(x_i, \dot{x}_i)\dot{x}_i = \tau_i + Y_i(t, x_i, \dot{x}_i)\theta_i \quad i = 1, \cdots, N. \tag{6.19}$$

We make the following assumption on the boundedness of $Y_i(t,x_i,\dot{x}_i)$:

Assumption 3 *If $\dot{x}_i$ is bounded, $Y_i(t,x_i,\dot{x}_i)$ is bounded for all $t \geq 0$ and $\dot{x}_i \in \mathbb{R}^p$.* □

The parameterization (6.18) encompasses a broad class of uncertainties. For example, when $Y_i(t,x_i,\dot{x}_i) = I_p$, (6.18) models constant disturbances. From standard result in the robotics literature [143, 105], (6.18) can represent the effects due to unknown constant parameters in $M_i(x_i)$ and $C_i(x_i,\dot{x}_i)$ in (6.5). To see this, we recall that both $M_i(x_i)$ and $C_i(\dot{x}_i,x_i)$ are linear in these constant parameters [70, 105]. Suppose that $M_i(x_i)$ and $C_i(x_i,\dot{x}_i)$ are estimated by $\hat{M}_i(x_i)$ and $\hat{C}_i(x_i,\dot{x}_i)$ whose parameters are the estimates of the unknown constant parameters in $M_i(x_i)$ and $C_i(x_i,\dot{x}_i)$. Instead of (6.5), agent i implements

$$\begin{aligned}\tau_i &= -K_i(\dot{x}_i - v(t)) + \hat{M}_i(x_i)\dot{v}(t) + \hat{C}_i(\dot{x}_i,x_i)v(t) + u_i \qquad (6.20)\\ &= -K_i(\dot{x}_i - v(t)) + M_i(x_i)\dot{v}(t) + C_i(\dot{x}_i,x_i)v(t) + u_i\\ &+ (\hat{M}_i(x_i) - M_i(x_i))\dot{v}(t) + (\hat{C}_i(\dot{x}_i,x_i) - C_i(\dot{x}_i,x_i))v(t). \qquad (6.21)\end{aligned}$$

Since $M_i(x_i)$ and $C_i(\dot{x}_i,x_i)$ are linear in robot parameters, the last two terms in (6.21) satisfy

$$(\hat{M}_i(x_i) - M_i(x_i))\dot{v}(t) + (\hat{C}_i(\dot{x}_i,x_i) - C_i(\dot{x}_i,x_i))v(t) = Y_i(\dot{v}(t),v(t),x_i,\dot{x}_i)\theta_i \qquad (6.22)$$

for some $Y_i(\dot{v}(t),v(t),x_i,\dot{x}_i)$ and θ_i. The closed-loop system of (6.1) and (6.21) then takes the form (6.19).

The next example uses the double integrator model (2.16) as a special case and illustrates the effects of the uncertainty $d_i(t,x_i,\dot{x}_i)$.

Example 6.1. A special case of the Euler Lagrange system is given by the double integrator model (2.16), where $M_i(x_i) = m_i I_p$, $C_i(x_i,\dot{x}_i) = \mathbf{0}_p$. In this case, (6.5) reduces to (2.17), which, together with (2.16), leads to

$$m_i\ddot{x}_i = -k_i(\dot{x}_i - v(t)) + m_i\dot{v}(t) + u_i. \qquad (6.23)$$

Suppose that agent i has only the nominal value of m_i, denoted by $\hat{m}_i$, for implementation and is subject to a constant disturbance $\theta_i^1 \in \mathbb{R}^p$ and viscous damping of the form $\theta_i^2\dot{x}_i$, where $\theta_i^2 \in \mathbb{R}$ is a constant. Then, the closed-loop dynamics of agent i become

$$\begin{aligned}m_i\ddot{x}_i &= -k_i(\dot{x}_i - v(t)) + \hat{m}_i\dot{v}(t) + u_i + \theta_i^1 + \theta_i^2\dot{x}_i \qquad (6.24)\\ &= -k_i(\dot{x}_i - v(t)) + m_i\dot{v}(t) + u_i + \underbrace{Y_i(t,x_i,\dot{x}_i)\theta_i}_{d_i(t,x_i,\dot{x}_i)} \qquad (6.25)\end{aligned}$$

where

$$Y_i(t,x_i,\dot{x}_i) = [I_p \;\; \dot{x}_i \;\; \dot{v}(t)] \qquad (6.26)$$

and

$$\theta_i = [(\theta_i^1)^T \;\; \theta_i^2 \;\; \hat{\theta}_i^3]^T = [(\theta_i^1)^T \;\; \theta_i^2 \;\; \hat{m}_i - m_i]^T. \qquad (6.27)$$

We see that $d_i(t,x_i,\dot{x}_i)$ encompasses the uncertainty in the model parameter, unknown viscous damping and constant disturbance.

We now perform a numerical simulation to see the effects due to $d_i(t,x_i,\dot{x}_i)$. Consider a group of four agents with $p = 1$. We choose $m_i = 1$ and $k_i = 2$, $\forall i$. The nominal values for m_i are set to $\hat{m}_1 = 0.9$, $\hat{m}_2 = 0.8$, $\hat{m}_3 = 1.1$ and $\hat{m}_4 = 1.2$. The viscous damping coefficient θ_i^2 is set to -0.05 while the constant disturbance θ_i^1 is chosen as 0.5, $\forall i$. We choose $v(t) = \sin(t)$. The initial values for x_i and $\dot{x}_i$ are all zero. The incidence matrix D is given by

$$D = \begin{pmatrix} 0 & 0 & -1 \\ 1 & 0 & 0 \\ -1 & 1 & 0 \\ 0 & -1 & 1 \end{pmatrix} \tag{6.28}$$

which means that only agents 2 and 3, agents 3 and 4 and agents 4 and 1 are neighbors. The weight matrix Δ is set to I_3.

The simulation result in Fig. 6.2 shows that the agreement objective fails due to the presence of $d_i(t,x_i,\dot{x}_i)$. Similarly, the tracking of the reference velocity is also not achieved, as illustrated in Fig. 6.3. □

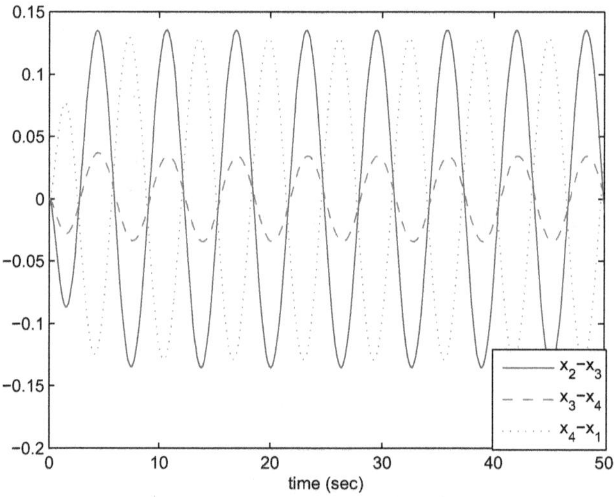

Fig. 6.2 In Example 6.1, the agreement of x_i's is not achieved because of the uncertainty $d_i(t,x_i,\dot{x}_i)$'s.

As shown in the above example, the uncertain system (6.19) with the nominal feedback law (6.5) cannot achieve the agreement of x_i's and objective A1. To compensate for the uncertainty $d_i(t,x_i,\dot{x}_i)$, we denote by $\hat{\theta}_i$ the estimate of θ_i and modify the control law (6.5) to

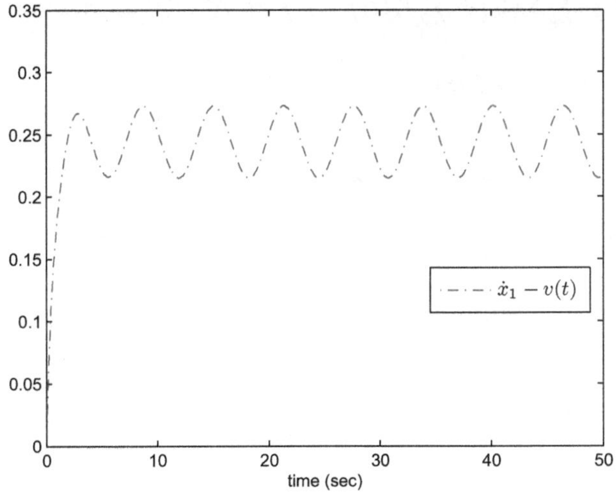

Fig. 6.3 In Example 6.1, $\dot{x}_i$'s do not track $v(t)$ because of the uncertainty $d_i(t, x_i, \dot{x}_i)$'s. For clarity, we only show $\dot{x}_1 - v(t)$ in this figure.

$$\tau_i = -K_i(\dot{x}_i - v(t)) + M_i(x_i)\dot{v}(t) + C_i(\dot{x}_i, x_i)v(t) + u_i - Y_i(t, x_i, \dot{x}_i)\hat{\theta}_i. \tag{6.29}$$

The goal in the next sections is to design the update law for $\hat{\theta}_i$ such that the control law in (6.29) guarantees the agreement of x_i's and objective A1.

6.4 A Preliminary Adaptive Design

We now present a preliminary adaptive design based on the storage functions $S_i(\xi_i)$ and $P_k(z_k)$ in (2.35). Let

$$\tilde{\theta}_i = \hat{\theta}_i - \theta_i \tag{6.30}$$

and

$$\tilde{\theta} = [\tilde{\theta}_1^T, \cdots, \tilde{\theta}_N^T]^T. \tag{6.31}$$

Note that (6.19) and (6.29) give rise to

$$\dot{x}_i = y_i + v(t), \tag{6.32}$$

$$\mathscr{H}_i : \begin{cases} M_i(x_i)\dot{\xi}_i = -C_i(x_i, \dot{x}_i)\xi_i - K_i\xi_i + u_i - Y_i(t, x_i, \dot{x}_i)\tilde{\theta}_i \\ y_i = \xi_i. \end{cases} \tag{6.33}$$

Consider the Lyapunov function

$$V = \sum_{i=1}^{N} S_i(\xi_i) + \sum_{k=1}^{\ell} P_k(z_k) + \frac{1}{2}\sum_{i=1}^{N} \tilde{\theta}_i^T \Lambda_i^{-1} \tilde{\theta}_i \quad \Lambda_i = \Lambda_i^T > 0 \tag{6.34}$$

whose time derivative is given by

$$\dot{V} = -\sum_{i=1}^{N} \xi_i^T K_i \xi_i - \sum_{i=1}^{N} \xi_i^T Y_i(t, x_i, \dot{x}_i)\tilde{\theta}_i + \tilde{\theta}_i^T \Lambda_i^{-1} \dot{\tilde{\theta}}_i. \tag{6.35}$$

If we design the update law for $\hat{\theta}_i$ as

$$\dot{\hat{\theta}}_i = \Lambda_i Y_i(t, x_i, \dot{x}_i)^T \xi_i \quad \Lambda_i = \Lambda_i^T > 0 \tag{6.36}$$

we obtain from (6.30) and (6.35) that

$$\dot{V} = -\sum_{i=1}^{N} \xi_i^T K_i \xi_i \leq 0 \tag{6.37}$$

which proves stability of the origin of $(\xi, z, \tilde{\theta})$.

Although the adaptation law (6.36) guarantees closed-loop stability, it does not ensure global convergence of $u \to 0$, i.e., the agreement of x_i's. To see this, let us look at the case where $Y_i(t, x_i, \dot{x}_i) = I_p$. We note that the equilibrium of (6.33) and (6.36) is given by $\xi_i = 0$ and $u_i = \tilde{\theta}_i$, which means that equilibria where $u \neq 0$ exist. Thus, $u \to 0$ cannot be achieved globally. The next numerical example confirms this.

Example 6.2. We consider Example 6.1 with only constant disturbance θ_i, which means in (6.25) $Y_i(t, x_i, \dot{x}_i) = I_p$. According to (6.29) and (6.36), each agent implements

$$\tau_i = -K_i(\dot{x}_i - v(t)) + m_i \dot{v}(t) + u_i - \hat{\theta}_i \tag{6.38}$$

where $\hat{\theta}_i$ is update by

$$\dot{\hat{\theta}}_i = \Lambda_i(\dot{x}_i - v(t)). \tag{6.39}$$

We take $\Lambda_i = 1$ and set the initial conditions of $\hat{\theta}_i$'s to $\hat{\theta}_1(0) = 0.1$, $\hat{\theta}_2(0) = 1.1$, $\hat{\theta}_3(0) = 0.3$ and $\hat{\theta}_4(0) = 0.7$. Fig. 6.4 shows that agreement is not achieved. □

One way to achieve $u \to 0$ globally is to ensure that the time derivative of the Lyapunov function contains a negative definite term of u. We next present two ways to achieve this: In the first approach, we employ the "*adding cross term*" technique [138, 144] and find another Lyapunov function for the nominal design (6.5) that yields a negative definite derivative in terms of ξ and u. An adaptive design based on this Lyapunov function guarantees $u \to 0$ and thus $z \to 0$; In the second approach we explore the design flexibility of the passivity framework and modify the nominal design (6.5) by feeding forward u_i to the velocity error $\dot{x}_i - v(t)$. As we will illustrate, this feedthrough term results in a closed-loop system similar to the celebrated *Slotine-Li* type system [121] for single robot tracking control. This approach yields better performance in convergence speed than the first approach but requires $\dot{u}$ information available for implementation.

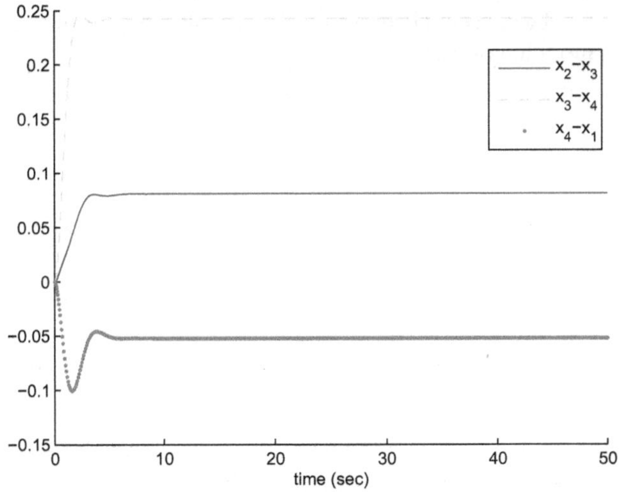

Fig. 6.4 In the presence of uncertainty, if we initialize $\hat{\theta}_1(0) = 0.1$, $\hat{\theta}_2(0) = 1.1$, $\hat{\theta}_3(0) = 0.3$ and $\hat{\theta}_4(0) = 0.7$, the agreement of x_i's is not achieved by (6.29) and (6.36).

6.5 Design 1: Adding a Cross Term

To state our result, we introduce the notation

$$M = \mathrm{diag}\{M_1(x_1), \cdots, M_N(x_N)\} \quad C = \mathrm{diag}\{C_1(x_1, \dot{x}_1), \cdots, C_N(x_N, \dot{x}_N)\}, \tag{6.40}$$

$$F_c = \mathrm{diag}\{f_{c1}, \cdots, f_{cN}\} \quad K = \mathrm{diag}\{K_1, \cdots, K_N\}, \tag{6.41}$$

$$Y = \mathrm{diag}\{Y_1(t, x_1, \dot{x}_1), \cdots, Y_N(t, x_N, \dot{x}_N)\} \tag{6.42}$$

and note from (6.33) that

$$M\dot{\xi} = -C\xi - K\xi + u - Y\tilde{\theta} \tag{6.43}$$

$$y = \xi = \dot{x} - 1_N \otimes v(t). \tag{6.44}$$

We now introduce the storage function

$$V_c = \frac{1}{2}\xi^T M \xi + \frac{1}{2} z^T (\Delta \otimes I_p) z - \varepsilon \xi^T M Q(u) u, \quad \varepsilon > 0 \tag{6.45}$$

where

$$Q(u) = \mathrm{diag}\left\{ \frac{1}{1 + 2|u_1|^2} I_p, \cdots, \frac{1}{1 + 2|u_N|^2} I_p \right\}. \tag{6.46}$$

The cross term $\varepsilon \xi^T M Q(u) u$ is added to create a negative definite term of u in $\dot{V}_c$ for sufficiently small ε. Proposition 6.1 below shows that this storage function guarantees passivity with respect to a modified output, namely $\dot{x} - 1_N \otimes v(t) - \varepsilon Q(u)u$ instead of $\xi = \dot{x} - 1_N \otimes v(t)$. Theorem 6.2 then replaces ξ in the adaptation law (6.36) with this modified output.

Proposition 6.1. *Let $Q(u)$ be defined as in (6.46) and let*

$$\varepsilon < \min\left(\frac{\lambda_{\min}(K)}{2\lambda_{\max}(M)\lambda_{\max}(L_\Delta) + \frac{\sqrt{N}}{2\sqrt{2}}\lambda_{\max}(F_c) + \frac{1}{4}\left(\mu_v \lambda_{\max}(F_c) + \lambda_{\max}(K)\right)^2},\right.$$
$$\left.\frac{\sqrt{\lambda_{\min}(M)\lambda_{\min}(\Delta)}}{\lambda_{\max}(M)\lambda_{\max}(D\Delta)}\right). \tag{6.47}$$

Then the dynamics in (6.43)-(6.44) with u defined in (6.9) are strictly passive from $-Y\tilde{\theta}$ to $\dot{x} - 1_N \otimes v(t) - \varepsilon Q(u)u$. □

Proof. We first show that V_c in (6.45) is indeed a positive definite function of z and ξ. We note from the definition of $Q(u)$ in (6.46) that

$$||Q(u)|| \leq 1, \tag{6.48}$$

which allows us to bound the third term in V_c as

$$\begin{aligned} \left|\varepsilon \xi^T M Q(u) u\right| &= \varepsilon \left|\xi^T M Q(u)(D\Delta \otimes I_p) z\right| \\ &\leq \varepsilon ||M|| \cdot ||D\Delta|| \cdot |\xi| \cdot |z|. \end{aligned} \tag{6.49}$$

Because

$$\frac{1}{2}\xi^T M \xi \geq \frac{\lambda_{\min}(M)}{2}|\xi|^2 \tag{6.50}$$

and

$$\frac{1}{2}z^T(\Delta \otimes I_p)z \geq \frac{\lambda_{\min}(\Delta)}{2}|z|^2, \tag{6.51}$$

we conclude that V_c is positive definite if

$$0 < \varepsilon < \frac{\sqrt{\lambda_{\min}(M)\lambda_{\min}(\Delta)}}{\lambda_{\max}(M)\lambda_{\max}(D\Delta)} \tag{6.52}$$

which is indeed the case from (6.47).

The time derivative of V_c along the trajectories of (6.43)-(6.44) is given by

$$\dot{V}_c = -\xi^T K \xi - \xi^T Y \tilde{\theta} - \varepsilon (M\dot{\xi} + \dot{M}\xi)^T Q(u) u - \varepsilon \xi^T M \dot{Q}(u) u - \varepsilon \xi^T M Q(u) \dot{u}. \tag{6.53}$$

Using (6.33) and the skew symmetry property (1.35), we further expand the third term in $\dot{V}_c$ as

$$\varepsilon (M\dot{\xi} + \dot{M}\xi)^T Q(u) u = \varepsilon\left[(C^T\xi - K\xi)^T Q(u) u + u^T Q(u) u - (Y\tilde{\theta})^T Q(u) u\right]. \tag{6.54}$$

We now consider the last two terms in (6.53). Note from (6.46) that

$$\dot{Q}(u)u = \text{diag}\left\{-\frac{4u_1u_1^T}{(1+2|u_1|^2)^2}, \cdots, -\frac{4u_Nu_N^T}{(1+2|u_N|^2)^2}\right\}\dot{u}. \tag{6.55}$$

Since

$$\left\|\frac{4u_iu_i^T}{(1+2|u_i|^2)^2}\right\| = \frac{4|u_i|^2}{(1+2|u_i|^2)^2} \leq \frac{4|u_i|^2}{1+4|u_i|^4} \leq 1 \tag{6.56}$$

we obtain

$$|\dot{Q}(u)u| \leq |\dot{u}|. \tag{6.57}$$

Using (6.9), (6.32) and (6.33), we get

$$\dot{u} = -(L_\Delta \otimes I_p)(\xi + 1_N \otimes v(t)) = -(L_\Delta \otimes I_p)\xi. \tag{6.58}$$

From (6.57) and (6.58), we bound the last two terms in (6.53) as

$$|\varepsilon\xi^T M\dot{Q}(u)u| \leq \varepsilon\lambda_{\max}(M)\lambda_{\max}(L_\Delta)|\xi|^2 \tag{6.59}$$
$$|\varepsilon\xi^T MQ(u)\dot{u}| \leq \varepsilon\lambda_{\max}(M)\lambda_{\max}(L_\Delta)|\xi|^2. \tag{6.60}$$

In particular, the last inequality (6.60) follows due to (6.48).

We now rewrite (6.53) from (6.48), (6.54), (6.59) and (6.60) as

$$\begin{aligned}\dot{V}_c \leq & -(\lambda_{\min}(K) - 2\varepsilon\lambda_{\max}(M)\lambda_{\max}(L_\Delta))|\xi|^2 - \varepsilon u^T Q(u)u + (\xi - \varepsilon Q(u)u)(-Y\tilde{\theta}) \\ & - \varepsilon(C^T\xi - K\xi)^T Q(u)u \qquad (6.61)\\ \leq & -(\lambda_{\min}(K) - 2\varepsilon\lambda_{\max}(M)\lambda_{\max}(L_\Delta))|\xi|^2 - \varepsilon|Q(u)u|^2 + (\xi - \varepsilon Q(u)u)(-Y\tilde{\theta}) \\ & + \varepsilon\lambda_{\max}(K)|\xi||u| + \varepsilon||C||\cdot|\xi|\cdot|Q(u)u| \qquad (6.62)\end{aligned}$$

where we used

$$u^T Q(u)u \geq u^T Q(u)^T Q(u)u \tag{6.63}$$

which follows from (6.46). Property 6.2, together with (6.2) and (6.32), implies that

$$||C|| \leq \lambda_{\max}(F_c)|\dot{x}| \leq \lambda_{\max}(F_c)|\xi| + \lambda_{\max}(F_c)\mu_v \tag{6.64}$$

which allows us to bound the last term in (6.62) as

$$\varepsilon||C||\cdot|\xi|\cdot|Q(u)u| \leq \frac{\sqrt{N}}{2\sqrt{2}}\varepsilon\lambda_{\max}(F_c)|\xi|^2 + \varepsilon\lambda_{\max}(F_c)\mu_v|\xi||u| \tag{6.65}$$

where we also used (6.48) and

$$|Q(u)u| \leq \frac{\sqrt{N}}{2\sqrt{2}}. \tag{6.66}$$

To see (6.66), we note from (6.46) that

$$|Q(u)u| = \sqrt{u^T Q(u)^T Q(u)u} = \sqrt{\sum_{i=1}^{N}\frac{|u_i|^2}{(1+2|u_i|^2)^2}} \leq \frac{\sqrt{N}}{2\sqrt{2}} \tag{6.67}$$

since

$$\frac{|u_i|^2}{(1+2|u_i|^2)^2} \leq \frac{1}{8}. \tag{6.68}$$

Then, $\dot{V}_c$ in (6.62) becomes

$$\dot{V}_c = -\begin{pmatrix} |\xi| \\ |Q(u)u| \end{pmatrix}^T \underbrace{\begin{pmatrix} a & b \\ b & \varepsilon \end{pmatrix}}_{P_c} \begin{pmatrix} |\xi| \\ |Q(u)u| \end{pmatrix} + (\xi - \varepsilon Q(u)u)^T(-Y\tilde{\theta}) \tag{6.69}$$

where

$$a = \lambda_{\min}(K) - 2\varepsilon\lambda_{\max}(M)\lambda_{\max}(L_\Delta) - \frac{\sqrt{N}}{2\sqrt{2}}\varepsilon\lambda_{\max}(F_c) \tag{6.70}$$

and

$$b = -\frac{1}{2}\varepsilon(\lambda_{\max}(F_c)\mu_v + \lambda_{\max}(K)). \tag{6.71}$$

Because

$$\varepsilon < \frac{\lambda_{\min}(K)}{2\lambda_{\max}(M)\lambda_{\max}(L_\Delta) + \frac{\sqrt{N}}{2\sqrt{2}}\lambda_{\max}(F_c) + \frac{1}{4}(\mu_v\lambda_{\max}(F_c) + \lambda_{\max}(K))^2}, \tag{6.72}$$

$a > 0$ and $a\varepsilon > b^2$, we conclude that P_c is positive definite. Noting $\xi = \dot{x} - 1_N \otimes v(t)$ in (6.44), we then obtain from (6.69) the strict passivity from $-Y\tilde{\theta}$ to $\dot{x} - 1_N \otimes v(t) - \varepsilon Q(u)u$. □

It is possible to choose the normalization matrix $Q(u)$ in (6.46) differently. For example, following [68], we can choose $Q(u) = \text{diag}\left\{\frac{1}{1+|u_1|}I_p, \cdots, \frac{1}{1+|u_N|}I_p\right\}$, which is bounded and ensures the boundedness of $Q(u)u$. If we were to set $Q(u)$ to identity matrix, we could not obtain the first bound in the right hand side of (6.65).

With the strict passivity property in Proposition 6.1, Theorem 6.2 below proves $\xi \to 0$ and $u \to 0$ if agent i implements the feedback law (6.29) with $\hat{\theta}_i$ updated by

$$\dot{\hat{\theta}}_i = \Lambda_i Y_i(t, x_i, \dot{x}_i)^T(\dot{x}_i - v(t) - \varepsilon\frac{u_i}{1+2|u_i|^2}), \quad \Lambda_i = \Lambda_i^T > 0. \tag{6.73}$$

Theorem 6.2. *Consider a group of N agents modeled as (6.19). Suppose that each agent implements (6.29) where u_i is as in (6.9) and $\hat{\theta}_i$ is updated by (6.73) with a positive ε satisfying (6.47). Then, the origin of $(\xi, z, \tilde{\theta})$ is stable and the trajectories of $(\xi, z, \tilde{\theta})$ are bounded. Moreover, $|\dot{x}_i - v(t)| \to 0$, $\forall i$ and $|x_i - x_j| \to 0$, $\forall i, j$.* □

Proof. The closed-loop system of (6.19), (6.29) and (6.73) is given by (6.43), (6.44) and

$$\dot{\tilde{\theta}}_i = \dot{\hat{\theta}}_i = \Lambda_i Y_i(t, x_i, \dot{x}_i)^T(\xi_i - \varepsilon\frac{u_i}{1+2|u_i|^2}). \tag{6.74}$$

We take

$$V_1 = V_c + \frac{1}{2}\sum_{i=1}^{N}\tilde{\theta}_i^T\Lambda_i^{-1}\tilde{\theta}_i \tag{6.75}$$

as a Lyapunov function, where V_c is in (6.45). Using Proposition 6.1 and (6.74), we obtain

$$\dot{V}_1 = -\begin{pmatrix} |\xi| \\ |Q(u)u| \end{pmatrix}^T P_c \begin{pmatrix} |\xi| \\ |Q(u)u| \end{pmatrix} \leq 0, \tag{6.76}$$

which implies stability of the origin of $(\xi, z, \tilde{\theta})$ and the boundedness of the signals $(\xi, z, \tilde{\theta})$. We further note from Property 6.1 and 6.2, Assumption 3 and (6.33) that $\dot{\xi}$ is bounded. Since $\dot{\xi}$ is bounded, $\dot{u}$ in (6.58) is bounded and thus $Q(u)\dot{u}$ is bounded. Using (6.57), we conclude that $\ddot{V}_1$ is bounded. We then apply Theorem B.5 and obtain $\xi \to 0$ and $Q(u)u \to 0$. It follows from (6.32) and (6.9) that $|\dot{x}_i - v(t)| \to 0$ and $|x_i - x_j| \to 0$, $\forall i, j$. □

Since the parameter ε is implemented in the update law (6.73), it must be chosen sufficiently small. To remove the dependence of ε on L_Δ, we can bound $||L_\Delta||$ by $N||\Delta||$. Given M, K, F_c and N, there is a permissible range of ε, which in turn determines the convergence rate.

Because ξ and u both converge to zero, we note from (6.33) and (6.73) that $Y_i(t, x_i, \dot{x}_i)\tilde{\theta}_i \to 0$ and $\dot{\tilde{\theta}}_i \to 0$. If, in addition, $Y_i(t, x_i, \dot{x}_i)^T$ satisfies the PE condition in (4.23), we further conclude from Lemma 4.1 that $\tilde{\theta}_i \to 0$, i.e., $\hat{\theta}_i \to \theta_i$. We illustrate this parameter convergence in the following example.

Example 6.3 (Example 6.1 continued).

We apply (6.29) and (6.73) to Example 6.1 and obtain the closed-loop system for agent i as

$$m_i \ddot{x}_i = -k_i(\dot{x}_i - v(t)) + m_i \dot{v}(t) + u_i + Y_i(t, x_i, \dot{x}_i)(\hat{\theta}_i - \theta_i) \tag{6.77}$$

where $v(t) = \sin(t)$, $Y_i(t, x_i, \dot{x}_i)$ is in (6.26) and θ_i is in (6.27).

In the simulation, we first take $\varepsilon = 0.15$ and randomly choose the initial conditions for $\hat{\theta}_i$. The update law for $\hat{\theta}_i$ is in (6.73) with Λ_i set to 2. All other initial conditions are the same as in Example 6.1. The simulation results in Figs. 6.5-6.6 show that the differences between x_i's and the tracking error $\dot{x}_i - v(t)$ converge to zero. With the same initial conditions, we next choose $\varepsilon = 0.05$. The agreement of x_i's is shown in Fig. 6.7, where the convergence is slower than the case of $\varepsilon = 0.15$.

The estimates $\hat{\theta}_i^j$ for θ_i^j in (6.27), $j = 1, 2, 3$, are shown in Figs. 6.8-6.10. We note that all $\hat{\theta}_i^j$'s converge to their true values. This is because the regressor matrix $Y_i(t, x_i, \dot{x}_i)^T$ becomes PE as $|\dot{x}_i - \sin(t)| \to 0$. To see this, note that $Y_i(t, x_i, \dot{x}_i)$ in (6.26) converges to $[1 \ \sin(t) \ \cos(t)]$ as $|\dot{x}_i - \sin(t)| \to 0$. The regressor $[1 \ \sin(t) \ \cos(t)]^T$ is PE since $\forall t_o \geq 0$

$$\int_{t_o}^{t_o+\delta} [1 \ \sin(s) \ \cos(s)]^T [1 \ \sin(s) \ \cos(s)] ds \geq \alpha I_3 \tag{6.78}$$

is satisfied with $\delta = 2\pi$ and $\alpha = \pi$. Let $\tilde{Y}_i = Y_i(t, x_i, \dot{x}_i) - [1 \ \sin(t) \ \cos(t)]$. Since $Y_i(t, x_i, \dot{x}_i)$ is a continuous and bounded function of t, we drop the dependence of Y_i and $\tilde{Y}_i$ on x_i and $\dot{x}_i$ and define

$$\kappa = \max_{t \geq 0} |Y_i(t)|. \tag{6.79}$$

It then follows from (6.78) that

$$\int_{t_o}^{t_o+\delta} (Y_i(s) - \tilde{Y}_i(s))^T (Y_i(s) - \tilde{Y}_i(s)) ds \tag{6.80}$$

$$= \int_{t_o}^{t_o+\delta} Y_i(s)^T Y_i(s) + \tilde{Y}_i(s)^T \tilde{Y}_i(s) - \tilde{Y}_i(s)^T Y_i(s) - Y_i(s)^T \tilde{Y}_i(s) ds \geq \alpha I_3. \tag{6.81}$$

Because $|\tilde{Y}_i| \to 0$, there exists a finite time t_1 such that $\forall t \geq t_1$, $|\tilde{Y}_i| \leq c$ for some positive scalar c. Thus, evaluating (6.81) at any $t' \geq t_1$ leads to

$$\alpha I_3 \leq \int_{t'}^{t'+\delta} Y_i(s)^T Y_i(s) + \tilde{Y}_i(s)^T \tilde{Y}_i(s) - \tilde{Y}_i(s)^T Y_i(s) - Y_i(s)^T \tilde{Y}_i(s) ds \tag{6.82}$$

$$\leq \int_{t'}^{t'+\delta} Y_i(s)^T Y_i(s) ds + c^2 \delta I_3 + 2 \int_{t'}^{t'+\delta} c |Y_i(s)| ds I_3$$

$$\leq \int_{t'}^{t'+\delta} Y_i(s)^T Y_i(s) ds + (2c\kappa\delta + c^2\delta) I_3. \tag{6.83}$$

Therefore, we have

$$\int_{t_1}^{t_1+\delta} Y_i(s)^T Y_i(s) ds \geq (\alpha - 2c\kappa\delta - c^2\delta) I_3. \tag{6.84}$$

Then for sufficiently small c, i.e., sufficiently large t_1, $\alpha - 2c\kappa\delta - c^2\delta$ is positive and thus $Y_i(t)$ is PE starting from time t_1. □

6.6 Design 2: Feedforward of the External Feedback

We note from Design 1 that the strict passivity in Proposition 6.1 is important in compensating for the uncertainty $d_i(t, x_i, \dot{x}_i)$. In this section, we present another design that guarantees this strict passivity by feeding forward u_i to y_i in (6.33). In particular, we replace $y_i = \xi_i$ in (6.33) with

$$y_i = \xi_i + \Gamma_i u_i \tag{6.85}$$

where $\Gamma_i = \Gamma_i^T > 0$. This modification is achieved with a new feedback law

$$\tau_i = M_i(x_i)\Gamma_i \dot{u}_i + C_i(\dot{x}_i, x_i)\Gamma_i u_i + M_i(x_i)\dot{v}(t) + C(\dot{x}_i, x_i)v(t) - K_i(\dot{x}_i - v(t) - \Gamma_i u_i)$$
$$+ u_i - Y_i(t, x_i, \dot{x}_i)\hat{\theta}_i. \tag{6.86}$$

Note that (6.86) employs $\dot{u}$, which implies from (6.9) that relative velocity information must be available.

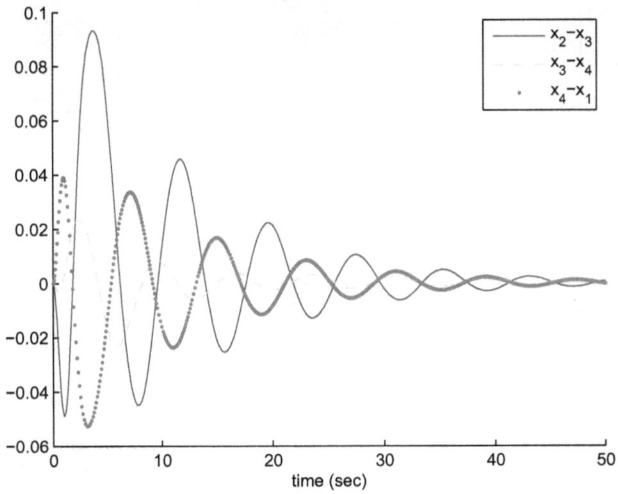

Fig. 6.5 Using the design (6.29) and (6.73) in Example 6.1, we achieve the agreement of x_i's.

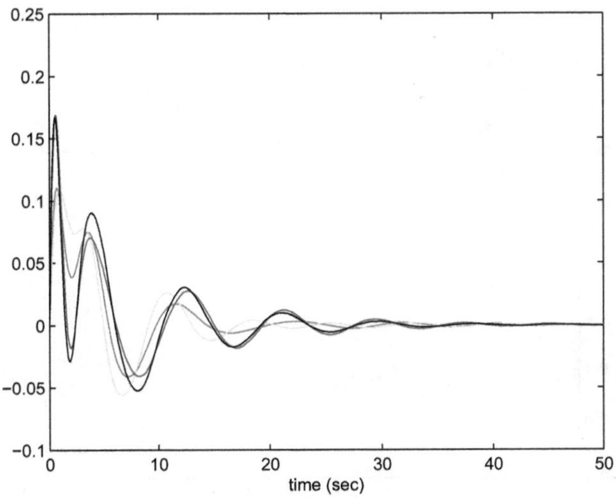

Fig. 6.6 The tracking errors $\dot{x}_i - v(t)$ converges to zero with the design (6.29) and (6.73).

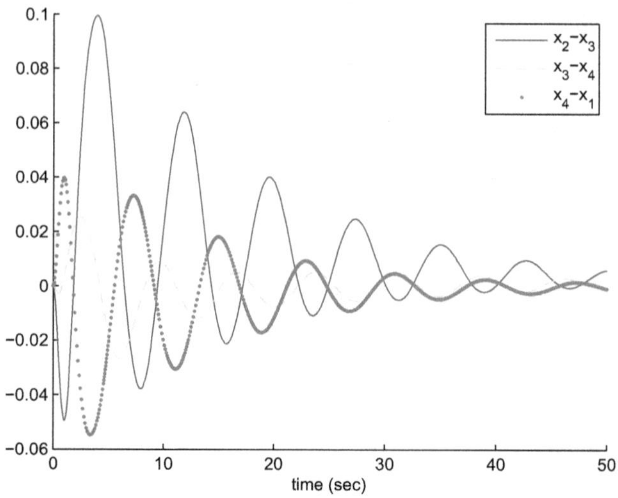

Fig. 6.7 The convergence rate is slower than that in Fig. 6.5 since we decrease ε from 0.15 to 0.05 in Example 6.3.

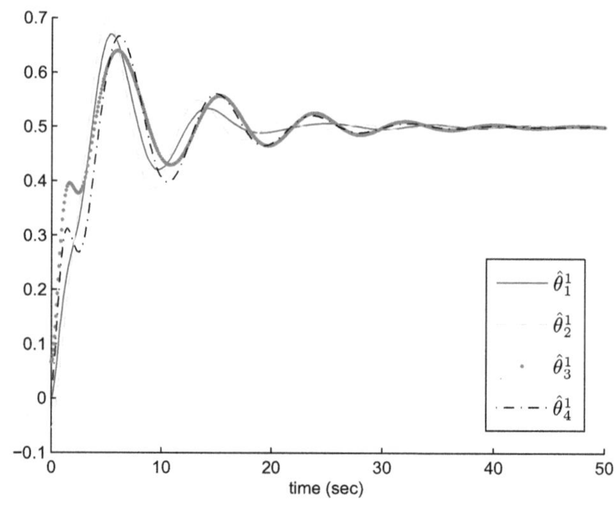

Fig. 6.8 The convergence of the estimates $\hat{\theta}_i^1$, $i = 1, \cdots, 4$.

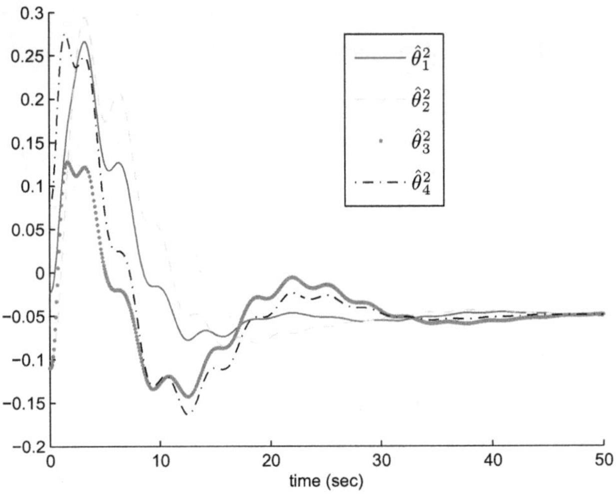

Fig. 6.9 The convergence of the estimates $\hat{\theta}_i^2$, $i = 1, \cdots, 4$.

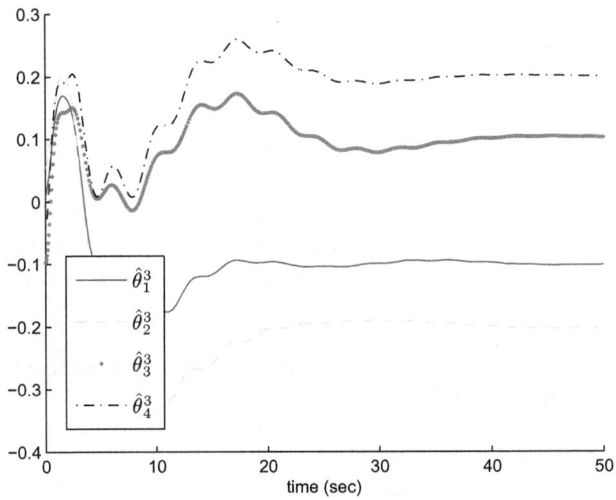

Fig. 6.10 The convergence of the estimates $\hat{\theta}_i^3$, $i = 1, \cdots, 4$.

We rewrite (6.32), (6.33) and (6.85) in a compact form

$$\dot{x} = \xi + \Gamma u + 1_N \otimes v(t) \tag{6.87}$$
$$M\dot{\xi} = -C\xi - K\xi + u - Y\tilde{\theta} \tag{6.88}$$

where $\Gamma = \text{diag}\{\Gamma_1, \cdots, \Gamma_N\}$. Proposition 6.2 below proves strict passivity from $-Y\tilde{\theta}$ to $\dot{x} - 1_N \otimes v(t) - \Gamma u$ with the storage function

$$V_s = \frac{1}{2}\xi^T M \xi + \frac{1}{2} z^T (\Delta \otimes I_p) z. \tag{6.89}$$

Proposition 6.2. *The dynamics in (6.87) and (6.88) are strictly passive from* $-Y\tilde{\theta}$ *to* $\dot{x} - 1_N \otimes v(t) - \Gamma u$. □

Proof. Consider the storage function (6.89). Its time derivative along the trajectories of (6.87) and (6.88) is given by

$$\dot{V}_s = -\xi^T K \xi - u^T \Gamma u - \xi^T Y \tilde{\theta}. \tag{6.90}$$

Noting $\xi = \dot{x} - 1_N \otimes v(t) - \Gamma u$ in (6.87), we obtain the required strict passivity result. □

We then propose the update law for $\hat{\theta}_i$ as

$$\dot{\hat{\theta}}_i = \Lambda_i Y_i(t, x_i, \dot{x}_i)^T (\dot{x}_i - v(t) - \Gamma_i u_i) \tag{6.91}$$

and prove the same stability and convergence results as in Theorem 6.2.

Theorem 6.3. *Consider a group of N agents modeled as (6.19). If each agent implements (6.86) where* u_i *is as in (6.9) and* $\hat{\theta}_i$ *is updated by (6.91), then the origin of* $(\xi, z, \tilde{\theta})$ *is stable and the trajectories of* $(\xi, z, \tilde{\theta})$ *are bounded. Moreover,* $|\dot{x}_i - v(t)| \to 0$ *and* $|x_i - x_j| \to 0$, $\forall i, j$. □

Proof. Using $V = V_s + \frac{1}{2}\sum_{i=1}^N \tilde{\theta}_i^T \Lambda_i^{-1} \tilde{\theta}_i$ as a Lyapunov function, we obtain from (6.90), (6.91) and (6.30) that

$$\dot{V} = -\sum_{i=1}^{N} (\xi_i^T K_i \xi_i + u_i^T \Gamma_i u_i) \leq 0. \tag{6.92}$$

The negative semidefinite derivative in (6.92) proves stability of the origin of $(\xi, z, \tilde{\theta})$ and the boundedness of the signals $(\xi, z, \tilde{\theta})$. The rest of the proof mimics the proof for Theorem 6.2. □

Theorem 6.3 relies on the $-u_i^T \Gamma_i u_i$ term in (6.92), which is achieved by feeding forward u_i to the velocity error y_i. The next example shows that with this feedforward term, the closed-loop system of (6.19) and (6.86) exhibits a form similar to the Slotine-Li type of controller in [121]. To illustrate this similarity in its basic form, we assume that $\tilde{\theta}_i = 0$ in this example.

Example 6.4. The Slotine-Li type controller [121] gives rise to a closed-loop system of the form

$$M_i(x_i)\ddot{x}_i + C_i(\dot{x}_i,x_i)\dot{x}_i = M_i(x_i)\Gamma_i(\dot{x}_i^d(t)-\dot{x}_i) + C_i(\dot{x}_i,x_i)\Gamma_i(x_i^d(t)-x_i) + M_i(x_i)\ddot{x}_i^d(t) + C(\dot{x}_i,x_i)\dot{x}_i^d(t) - K_i[\dot{x}_i - \dot{x}_i^d(t) + \Gamma_i(x_i - x_i^d(t))], \qquad (6.93)$$

which achieves tracking of $|\dot{x}_i - \dot{x}_i^d(t)| \to 0$ and $|x_i - x_i^d(t)| \to 0$. The closed-loop system of (6.19) and (6.86) is given by

$$M_i(x_i)\ddot{x}_i + C_i(\dot{x}_i,x_i)\dot{x}_i = M_i(x_i)\Gamma_i\dot{u}_i + C_i(\dot{x}_i,x_i)\Gamma_i u_i + M_i(x_i)\dot{v}(t) + C(\dot{x}_i,x_i)v(t) - K_i[\dot{x}_i - v(t) - \Gamma_i u_i] + u_i. \qquad (6.94)$$

Comparing (6.93) and (6.94), we note that u_i and $\dot{x}_i - v(t)$ in (6.94) are similar to $x_i - x_i^d(t)$ and $\dot{x}_i - \dot{x}_i^d(t)$ in (6.93), respectively. In fact, in the Slotine-Li controller, the tracking objective is represented in the absolute coordinate x_i while in our case, each agent's tracking objective is $u_i \to 0$, which is represented in the coordinate of the relative variables z_k. The last term u_i in (6.94) is used for interconnecting with the $\dot{z}$ dynamics. □

We now demonstrate that tuning the free parameter Γ_i can yield better performance in compensating for the uncertainty than the approach in Section 6.5. Note that this is at the cost of employing relative velocity information in (6.86).

Example 6.5 (Example 6.1 continued).

We now use Design 2 to compensate for the uncertainty in Example 6.1. Applying (6.86) to the double integrator model (2.16) and noting that only the nominal values for m_i are known, we obtain

$$\tau_i = \hat{m}_i\Gamma_i\dot{u}_i + \hat{m}_i\dot{v}(t) - k_i(\dot{x}_i - v(t) - \Gamma_i u_i) + u_i + \theta_i^1 + \theta_i^2\dot{x}_i - Y_i(t,x_i,\dot{x}_i)\hat{\theta}_i \qquad (6.95)$$

where the form of $Y_i(t,x_i,\dot{x}_i)$ is to be specified. We further rewrite (6.95) as

$$m_i\ddot{x}_i = m_i\Gamma_i\dot{u}_i + m_i\dot{v}(t) - k_i(\dot{x}_i - v(t) - \Gamma_i u_i) + u_i + Y_i(t,x_i,\dot{x}_i)\theta_i - Y_i(t,x_i,\dot{x}_i)\hat{\theta}_i \qquad (6.96)$$

where θ_i is as in (6.27) and $Y_i(t,x_i,\dot{x})$ is given by

$$Y_i(t,x_i,\dot{x}_i) = [1 \;\; \dot{x}_i \;\; \dot{v}(t) + \Gamma_i\dot{u}_i]. \qquad (6.97)$$

The update law for $\hat{\theta}_i$ is given by (6.91).

In the simulation, we choose $\Gamma_i = 0.15$ in (6.95), the same as ε in Example 6.3. All the other parameters are the same as in Example 6.3. The simulation result in Fig. 6.11 shows that the agreement of x_i's is achieved while Fig. 6.12 illustrates the tracking error $\dot{x}_i - v(t)$ converges to zero, $\forall i$. Note that the convergence in Figs. 6.11 and 6.12 is similar to that in Figs. 6.5 and 6.6.

Unlike ε in Example 6.3, Γ_i is a free parameter. We then increase Γ_i to improve the convergence. Fig. 6.13 shows the agreement result for $\Gamma_i = 1$ with the same initial conditions as the case of $\Gamma_i = 0.15$. The convergence is faster and the oscillations are smaller than those in Fig. 6.11, where $\Gamma_i = 0.15$. □

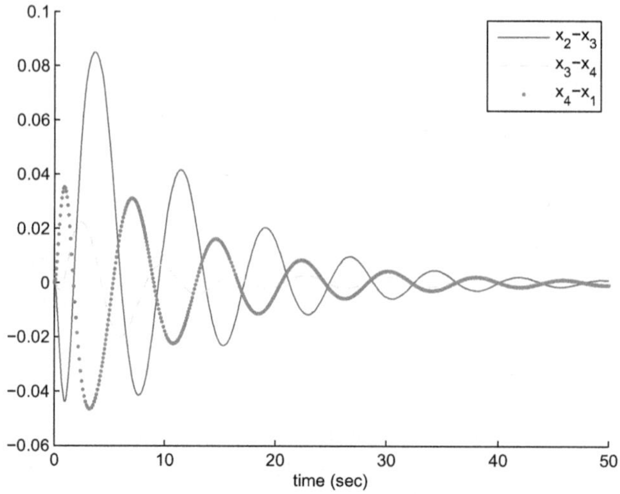

Fig. 6.11 Using the design (6.86) and (6.91) in Example 6.1, we achieve the agreement of x_i's.

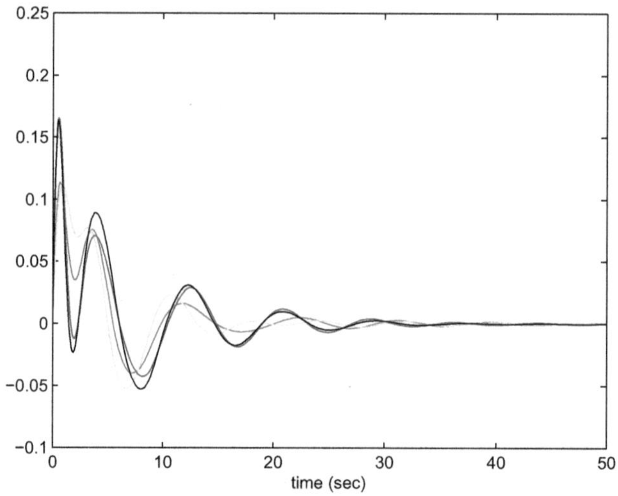

Fig. 6.12 The tracking error $\dot{x}_i - v(t)$ converges to zero with the design (6.86) and (6.91).

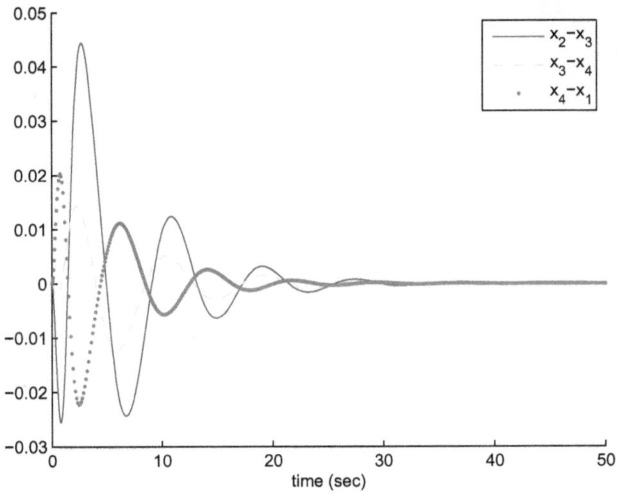

Fig. 6.13 The oscillation is smaller than that in Fig. 6.11 and the convergence is faster because we have increased Λ_i from 0.15 to 1 in Example 6.5.

6.7 Summary

In this chapter, we studied the agreement of multiple Euler Lagrange systems with parameterized uncertainty. We presented two approaches to compensating for the uncertainty. The first approach relies on the inherent robustness of Euler Lagrange systems and employs the "adding cross term" technique. The second approach feeds forward the external feedback and takes a form similar to the Slotine-Li type controller. The second approach achieves a faster convergence than the first approach. However, relative velocity information must be available to implement the second approach.

Chapter 7
Synchronized Path Following

7.1 Introduction

So far, we have focused on the formation control problem—maintaining the desired formation towards a common goal. For reconnaissance missions or maneuvering around obstacles, additional requirement of following a specified path may be required. The primary goal in *path following* problems is to design control laws that force the output of a system to follow a desired *path*. The secondary goal for the system is to obey a speed assignment along the path. In particular, we will consider the formulation where we parameterize the desired path by a path variable θ, and assign a desired speed to be achieved by $\dot{\theta}$ as $t \to \infty$. Unlike the classical tracking problem, where the speed assignment must be fulfilled for all $t \geq 0$, the path following problem offers flexibility to shape the transient behavior of $\dot{\theta}$.

In this chapter, we exploit this flexibility to synchronize the path variables for a group of path-following systems. When the path variables reach agreement, the group exhibits a desired formation. Note that this synchronized path following approach achieves the desired formation through coordinated tracking of individual trajectories. This is different from the formulations in position/distance-based formation control studied in Sections 2.6 and 2.7, where the desired formation is achieved by controlling relative configurations. To synchronize the path variables θ, we apply the agreement schemes in Section 2.5 and obtain two synchronized path following designs. In the first design, the agreement schemes in Section 2.5 and the path following systems constitute a cascaded system. Because the agreement schemes are globally asymptotically stable, we demonstrate the closed-loop stability of this cascaded system by showing that the path following systems are Input-to-State Stable (ISS). In the second design, we will represent the closed-loop system as the feedback interconnection of a dynamic block for path variable agreement and another block that incorporates the path following systems. Since the agreement block is passive, we prove closed-loop stability by showing the path following systems are also passive, thereby satisfying Structure 3 in Section 1.5.

7.2 Path-following Design and Synchronization

Consider a general system

$$\begin{aligned} \dot{x} &= f(x,u) \\ y &= h(x) \end{aligned} \tag{7.1}$$

where $x \in \mathbb{R}^n$ is the state vector, $y \in \mathbb{R}^m$ is the system output and $u \in \mathbb{R}^n$ is the control. The objective of the path-following problem is to force y to a prescribed feasible path $y_d(\theta)$ and to assign a feasible speed $v(t)$ to $\dot{\theta}$ on this path, where θ is a scalar. The parameterized path $y_d(\theta)$ is a geometric curve defined as

$$Y_d := \{y \in \mathbb{R}^m | \ \exists\theta \in \mathbb{R} \text{ such that } y = y_d(\theta)\} \tag{7.2}$$

where y_d is continuously parameterized by the path variable θ.

We assume that there exists a preliminary design that brings (7.1) to a closed-loop system of the form

$$\begin{aligned} \dot{z} &= F(x)z - g(t,x,\theta)\omega \\ \dot{\theta} &= v(t) - \omega \end{aligned} \tag{7.3}$$

where z is a set of new parameters that include the tracking error $y - y_d(\theta)$ and its derivatives, and ω is a feedback term to be designed such that the reference velocity $v(t)$ is achieved asymptotically, i.e.,

$$\omega \to 0 \text{ as } t \to \infty. \tag{7.4}$$

The matrix $F(x) \in \mathbb{R}^{n\times n}$ and the vector $g(t,x,\theta) \in \mathbb{R}^n$ depend on the control design. In particular, $F(x)$ satisfies

$$PF(x) + F^T(x)P \leq -I \tag{7.5}$$

for some matrix $P = P^T > 0$. In Section 7.4, we will demonstrate how to transform dynamics of marine vehicles to the form in (7.3) using backstepping. Preliminary design examples can also be found in [120].

Assumption 4 *For the system in (7.1), a feasible path $y_d(\theta)$ and uniformly bounded speed assignments $v(t)$ imply that the function $g(t,x,\theta)$ is uniformly bounded in its arguments.* □

We now consider a group of agents $i = 1,\cdots,N$, each controlled by an individual path-following design with a prescribed velocity $v(t)$, leading to the closed-loop system

$$\Sigma_i : \begin{cases} \dot{z}_i = F_i(x_i)z_i - g_i(t,x_i,\theta_i)\omega_i \\ \dot{\theta}_i = v(t) - \omega_i. \end{cases} \tag{7.6}$$

The goal of synchronized path-following is to design ω_i such that the path variables θ_i, $i = 1,\cdots,N$, reach agreement with (7.4) satisfied. The design of ω_i depends on variables of agent i and the path parameters from its neighboring agents. Therefore, only one scalar variable needs to be transmitted from each agent to its neighbors. For convenience, we introduce the concatenated vectors

$$\theta = [\theta_1, \cdots, \theta_N]^T \quad \omega = [\omega_1, \cdots, \omega_N]^T \tag{7.7}$$

$$z = [z_1, \cdots, z_N]^T \quad g = \mathrm{diag}\{g_1, \cdots, g_N\}. \tag{7.8}$$

7.3 Passivity-based Designs for Synchronization

7.3.1 Design 1: Without Path Error Feedback

A straightforward idea for synchronizing θ_i's in (7.6) is to apply the result from Corollary 2.1. To this end, we design the input ω_i as

$$\omega_i = \mathscr{H}_i\left\{\sum_{i=1}^{\ell} d_{ik}\psi_k(\tilde{\theta}_k)\right\}, \quad i = 1, \cdots, N \tag{7.9}$$

where $\mathscr{H}_i\{u_i\}$ satisfies (2.11)-(2.15) and $\tilde{\theta}_k$ is defined as

$$\tilde{\theta}_k := \sum_{l=1}^{N} d_{lk}\theta_l = \begin{cases} \theta_i - \theta_j & \text{if } k \in \mathscr{L}_i^+ \\ \theta_j - \theta_i & \text{if } k \in \mathscr{L}_i^-. \end{cases} \tag{7.10}$$

The definition of $\tilde{\theta}_k$ is similar to z_k defined in (2.2). Therefore, similarly to (2.6), we have

$$\tilde{\theta} = D^T\theta \tag{7.11}$$

where $\tilde{\theta} = [\tilde{\theta}_1, \cdots, \tilde{\theta}_\ell]^T$. The nonlinearity $\psi_k(\cdot)$ satisfies (2.27)-(2.31) and (2.46). Following Corollary 2.1, we conclude that (7.9) ensures global asymptotic stability of the origin of $(\xi, \tilde{\theta})$, i.e., the agreement of θ_i's and (7.4) are achieved.

Does (7.9) also guarantee $z_i \to 0$, i.e., does each agent track its prescribed path? To answer this question, we note that the closed-loop system (7.6) and (7.9) becomes a cascaded system in Fig. 7.1. In Theorem 7.1 below we prove that Σ_i in (7.6) is Input-to-State Stable (ISS) with respect to ω_i. Stability of the closed-loop system then follows because a cascade of an ISS and a uniformly globally asymptotically stable system is uniformly globally asymptotically stable (see Theorem B.7 in Appendix B.6). Thus, each agent follows its desired path, that is, $z_i \to 0$.

Theorem 7.1. *Consider the dynamics of the ith agent in (7.6), $i = 1, \cdots, N$. Let the control ω_i in (7.6) be as (7.9), where $\mathscr{H}_i\{\cdot\}$ denotes the output at time t of a static or dynamic block satisfying (2.11)-(2.15). Then, (7.6) is ISS with respect to the input ω_i and the origin of $(\xi, z, \tilde{\theta})$ is uniformly globally asymptotically stable.* □

Proof. We only need to demonstrate the ISS property of Σ_i in (7.6). To do so, we define a storage function

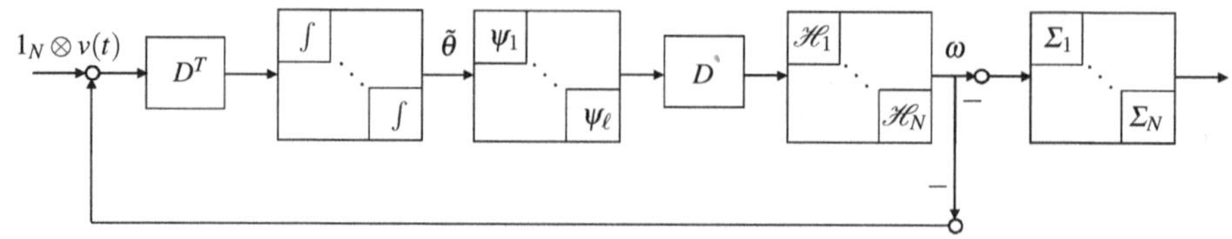

Fig. 7.1 Block diagram for the synchronized path following Design 1.

$$V_z = \sum_{i=1}^{N} z_i^T P_i z_i \tag{7.12}$$

where P_i is as in (7.5). Its time derivative along (7.6) is given by

$$\dot{V}_z \leq -\sum_{i=1}^{N} z_i^T z_i + z_i^T P_i g_i \omega_i \tag{7.13}$$

$$\leq -\sum_{i=1}^{N} |z_i|^2 + 2p_{im}\delta_{gi}|z_i||\omega_i| \tag{7.14}$$

where $p_{im} = ||P_i||$ and δ_{gi} is an upper bound on g_i due to Assumption 4. Furthermore, we get

$$|z_i| \geq \frac{2p_{im}\delta_{gi}|\omega_i|}{\varepsilon} \Longrightarrow \dot{V}_z \leq -\sum_{i=1}^{N}(1-\varepsilon)|z_i|^2, \tag{7.15}$$

where $0 < \varepsilon < 1$. Thus, it follows from Theorem B.8 in Appendix B.7 that the system is ISS from ω_i to z_i with $\rho(r) = \frac{2p_{im}\delta_{gi}r}{\varepsilon}$. Since the origin of $(\xi, \tilde{\theta})$ is globally asymptotically stable by the design in Corollary 2.1 and Σ_i is ISS with respect to ω_i, it follows from Theorem B.7 that the origin of $(\tilde{\theta}, \xi, z)$ is uniformly globally asymptotically stable. □

Remark 7.1. The cascaded structure of Design 1 in Fig. 7.1 also allows us to apply the adaptive designs in Chapters 3 and 4 when only one agent has the $v(t)$ information. Since the desired target set for $\tilde{\theta}$ is the origin, both adaptive designs in Corollary 3.1 and Theorem 4.2 ensure $\omega \to 0$ and the synchronization of θ_i's. Because the adaptive design cascaded to the Σ_i blocks is globally uniformly asymptotically stable and because the Σ_i blocks are proven to be ISS, it follows from Theorem B.7 again that the tracking error z converges to zero and the synchronized path following is achieved. □

7.3.2 Design 2: With Path Error Feedback

When the path error z_i is available for feedback design, we choose

$$\omega_i = \mathscr{H}_i\{\tilde{u}_i\} \tag{7.16}$$

where

$$\tilde{u}_i = 2z_i^T P_i g_i + \sum_{i=1}^{\ell} d_{ik}\psi_k(\tilde{\theta}_k). \tag{7.17}$$

The first term in (7.17) is the path error feedback that serves to improve convergence to the desired path while the second term is for the agreement of θ_i's.

If $\mathscr{H}_i$ is static, we restrict it to be of the form

$$\omega_i = h_i(\tilde{u}_i) \tag{7.18}$$

where the function $h_i(\cdot)$ is strictly passive as defined in Definition 1.3.

If $\mathscr{H}_i$ is dynamic, we assume that it is of the form

$$\dot{\xi}_i = f_i(\xi_i) + a_i(\xi_i)\tilde{u}_i \tag{7.19}$$

$$\omega_i = h_i(\xi_i) + b_i(\xi_i)\tilde{u}_i \tag{7.20}$$

where $f_i(\cdot)$, $a_i(\cdot)$, $h_i(\cdot)$, and $b_i(\cdot)$ are continuous functions such that $f_i(0) = 0$ and $h_i(0) = 0$. The main restriction of (7.19)-(7.20) is that it be passive with a C^2 positive definite and radially unbounded storage function $S_i(\xi_i)$ satisfying

$$\dot{S}_i \leq -W_i(\xi_i) + \tilde{u}_i\omega_i - c_i\tilde{u}_i^2, \quad c_i \geq 0 \tag{7.21}$$

for some positive definite function $W_i(\xi_i)$. Inequality (7.21) with $c_i > 0$ implies *strict input passivity*, which can be achieved only when (7.19)-(7.20) has a relative degree zero. Our asymptotic stability result allows $c_i = 0$ if (7.19)-(7.20) has well-defined relative degree one at $\xi_i = 0$, that is

$$b_i(\xi_i) = 0, \quad a_i(0) \neq 0, \quad \frac{\partial h_i(\xi_i)}{\partial \xi_i}\Big|_{\xi_i=0} \neq 0. \tag{7.22}$$

We thus make the following assumption:

Assumption 5 *If $c_i = 0$ in (7.21), then (7.22) holds.* □

With ω_i in (7.16), the closed-loop system is shown in Fig. 7.2, where

$$u_i = -\sum_{i=1}^{\ell} d_{ik}\psi_k(\tilde{\theta}_k), \quad u = [u_1, \cdots, u_N]^T \tag{7.23}$$

and

$$Y_i = 2z_i^T P_i g_i, \quad Y = [Y_1, \cdots, Y_N]^T. \tag{7.24}$$

We investigate stability properties of the closed-loop system by separating it into two blocks, $\mathscr{S}_1$ and $\mathscr{S}_2$ as in Fig. 7.2, and analyze passivity properties of each block. The following theorem shows the passivity from $-\omega$ to Y and thus the passivity of $\mathscr{S}_2$ from u to ω. Because $\mathscr{S}_1$ is also passive (see Fig. 2.2 and Theorem 2.1), we conclude closed-loop stability from Structure 3 in Section 1.5. We further prove

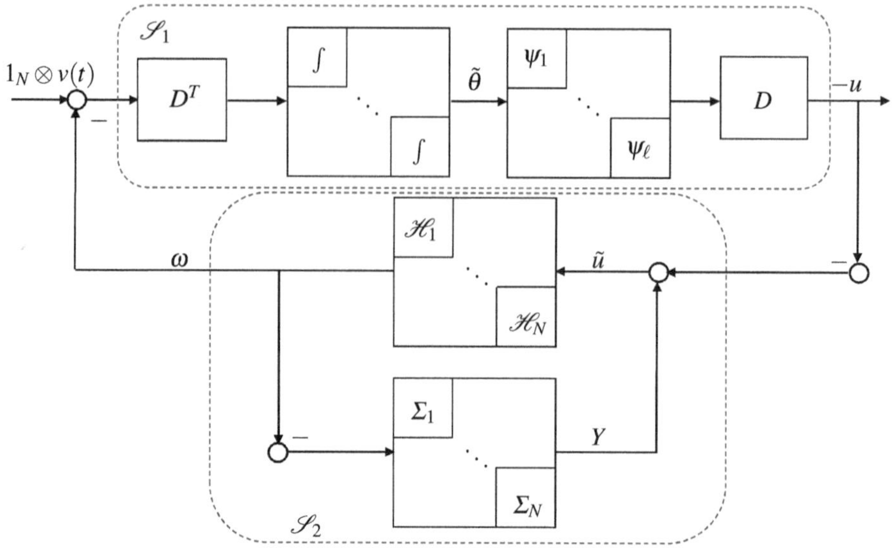

Fig. 7.2 Block diagram for synchronized path following Design 2.

the uniformly globally asymptotic stability of the origin of $(\tilde{\theta}, \xi, z)$ by applying the Nested Matrosov Theorem.

Theorem 7.2. *Consider the dynamics of the ith agent in (7.6), $i = 1, \cdots, N$. Let the control ω_i in (7.6) be as (7.16), where $\mathscr{H}_i$ satisfies (7.18)-(7.21). If $\mathscr{H}_i$ is dynamic, we assume that Assumption 5 holds. Then, Σ_i is strictly passive from $-\omega_i$ to Y_i. The origin of $(\xi, z, \tilde{\theta})$ is uniformly globally asymptotically stable.* □

Proof. We demonstrate the passivity of Σ_i by considering the storage function in (7.12), whose time derivative along the trajectories of z is given by

$$\dot{V}_z = -\sum_{i=1}^{N} (z_i^T z_i + Y_i^T \omega_i). \tag{7.25}$$

Thus, the passivity from $-\omega_i$ to Y_i is established.

We combine the results from Theorems 2.1 and 4.2 to prove the uniformly globally asymptotical stability of $(\xi, z, \tilde{\theta})$. We first note from Theorem 2.1 that the feedforward path $\mathscr{S}_1$ is passive from ω to u with the storage function

$$V_p = \sum_{k=1}^{\ell} P_k(\tilde{\theta}_k) \tag{7.26}$$

where $P_k(\cdot)$ is as in (2.27)-(2.31).

We next show the passivity of $\mathscr{S}_2$. Let $\mathscr{I}$ denote the set of indices $i = 1,\cdots,N$ for which $\mathscr{H}_i$ is a dynamic block as (7.19)-(7.20). Using the storage function $V_b(\xi)$ in (2.35) and (7.21), we obtain

$$\dot{V}_b(\xi) = \sum_{i\in\mathscr{I}} (-W_i(\xi_i) + \tilde{u}_i\omega_i) \tag{7.27}$$

$$\leq -\sum_{i\in\mathscr{I}} W_i(\xi_i) + \sum_{i=1}^{N} \tilde{u}_i\omega_i - \sum_{i\notin\mathscr{I}} \tilde{u}_i\omega_i. \tag{7.28}$$

We further use the Lyapunov function

$$V = V_b + V_p + V_z \tag{7.29}$$

whose time derivative is given by

$$\dot{V} = \sum_{i=1}^{N}(-z_i^T z_i - c_i\tilde{u}_i^2 + Y_i^T\omega_i + u_i^T\omega_i + \tilde{u}_i\omega_i) - \sum_{i\notin\mathscr{I}} \tilde{u}_i\omega_i - \sum_{i\in\mathscr{I}} W_i(\xi_i) \tag{7.30}$$

$$= -\sum_{i=1}^{N} z_i^T z_i - \sum_{i\notin\mathscr{I}} \tilde{u}_i\omega_i - \sum_{i\in\mathscr{I}} (W_i(\xi_i) + c_i\tilde{u}_i^2) \leq 0. \tag{7.31}$$

Finally, since the static blocks satisfy (1.25),

$$\sum_{i\notin\mathscr{I}} \tilde{u}_i\omega_i = \sum_{i\notin\mathscr{I}} \tilde{u}_i h_i(u_i) \geq 0. \tag{7.32}$$

We thus conclude that $\dot{V}$ is negative semidefinite and that the equilibrium $(z,\xi,\tilde{\theta}) = 0$ is uniformly stable.

To establish uniform asymptotic stability, we use the Nested Matrosov Theorem in Theorem B.5. Similarly to the proof of Theorem 4.2, we define an auxiliary function

$$V_2 = -\tilde{\theta}^T D^+ \Gamma \omega \tag{7.33}$$

where D^+ is the pseudo-inverse of the incidence matrix D and Γ is a diagonal matrix with entries

$$\Gamma_{ii} = \begin{cases} (L_{a_i}h_i(0))^{-1} & \text{if } i \in \mathscr{I} \\ 0 & \text{if } i \notin \mathscr{I}. \end{cases} \tag{7.34}$$

In particular $L_{a_i}h_i(0) := \frac{\partial h_i(\xi_i)}{\partial \xi_i}\big|_{\xi_i=0} a_i(0)$ is nonsingular and thus invertible because of the passivity of the ξ_i-subsystems in (7.19)-(7.20) and because of Proposition B.1 in Appendix B.4. To apply the Nested Matrosov Theorem, we denote by Y_1 the right hand side of (7.31) and claim that

$$Y_1 = 0 \Longrightarrow \dot{V}_2 := Y_2 \leq 0. \tag{7.35}$$

To see this, we note that $Y_1 = 0$ implies that $\xi = 0$ and $\omega = 0$, which mean that all terms in $\dot{V}_2$ vanish except

$$-\tilde{\theta}^T D^+ \Gamma \dot{\omega}\big|_{Y_1=0}. \tag{7.36}$$

Because $\dot{\omega}_i\big|_{\xi=0} = L_{a_i} h_i(0)\tilde{u}_i$ when $i \in \mathscr{I}$ and $c_i = 0$ and because $Y_1 = 0$ implies $\tilde{u}_i = 0$ for $i \notin \mathscr{I}$ or $c_i > 0$, we conclude from (7.34) that $\Gamma\dot{\omega}\big|_{Y_1=0} = \tilde{u}$ and obtain

$$-\tilde{\theta}^T D^+ \Gamma \dot{\omega}\big|_{Y_1=0} = -\tilde{\theta}^T D^+ \tilde{u}. \tag{7.37}$$

Since $Y_1 = 0$ implies $z = 0$, it follows from (7.17) and (7.23) that $\tilde{u}_i = -u_i$. Substituting (7.11) in (7.37) and using the property $DD^+D = D$, we conclude

$$Y_1 = 0 \Longrightarrow Y_2 = -\theta^T DD^+ D\psi(\tilde{\theta}) = -\tilde{\theta}^T \psi(\tilde{\theta}). \tag{7.38}$$

Since $\psi_k(\cdot)$ satisfies (2.46), $\tilde{\theta}^T \psi(\tilde{\theta})$ is positive definite in $\tilde{\theta}$. Thus, (7.38) proves the claim (7.35). It follows from (7.31) and (7.38) that $Y_1 = 0$ and $Y_2 = 0$ together imply $(z, \xi, \tilde{\theta}) = 0$. All assumptions of Theorem B.5 are satisfied and we conclude uniformly globally asymptotic stability of $(z, \xi, \tilde{\theta}) = 0$. □

7.4 Design Example

7.4.1 Agent Dynamics

Suppose that each agent is a fully actuated tugboat with three degrees of freedom, where the surge mode is decoupled from the sway and yaw mode due to port/starboard symmetry. The dynamic model of agent i, $i = 1, \cdots, N$, is given by [49]:

$$\dot{\eta}_i = R_i \nu_i \tag{7.39}$$

$$M_i \dot{\nu}_i + D_i(\nu_i)\nu_i = \tau_i \tag{7.40}$$

where $\eta_i = [x_i, y_i, \psi_i]^T$, $(x_i, y_i) \in \mathbb{R}^2$ is the position vector in the inertial frame E, ψ_i is the heading angle (yaw), and $\nu_i = [\nu_{i,1}, \nu_{i,2}, \nu_{i,3}]^T \in \mathbb{R}^3$ is the velocity vector in the body frame B. The model matrices M_i and $D_i(\nu_i)$ denote inertia, Coriolis plus centrifugal and damping, respectively, while $\tau_i \in \mathbb{R}^3$ is the generalized control forces and moments in the body frame. The matrix $R_i \in SO(3)$ is the attitude matrix of the tugboat with respect to the inertial frame (see Fig. 7.3). Given ψ_i, R_i can be written as

$$R_i = R_i(\psi_i) = \begin{pmatrix} \cos\psi_i & -\sin\psi_i & 0 \\ \sin\psi_i & \cos\psi_i & 0 \\ 0 & 0 & 1 \end{pmatrix}. \tag{7.41}$$

Recall that R_i satisfies $R_i^T R_i = I_3$, and

$$\dot{R}_i = R_i S \dot{\psi}_i = \dot{\psi}_i S R_i \tag{7.42}$$

where

$$S = \begin{pmatrix} 0 & -1 & 0 \\ 1 & 0 & 0 \\ 0 & 0 & 0 \end{pmatrix}. \tag{7.43}$$

The numerical values for (7.39)-(7.40) are taken from [60] as

$$M_i = \begin{pmatrix} 33.8 & 1.0948 & 0 \\ 1.0948 & 2.764 & 0 \\ 0 & 0 & 23.8 \end{pmatrix} \tag{7.44}$$

and

$$C_i(v_i) = \begin{pmatrix} 7 & 0.1 & 0 \\ 0.1 & 0.5 & 0 \\ 0 & 0 & 2 \end{pmatrix}. \tag{7.45}$$

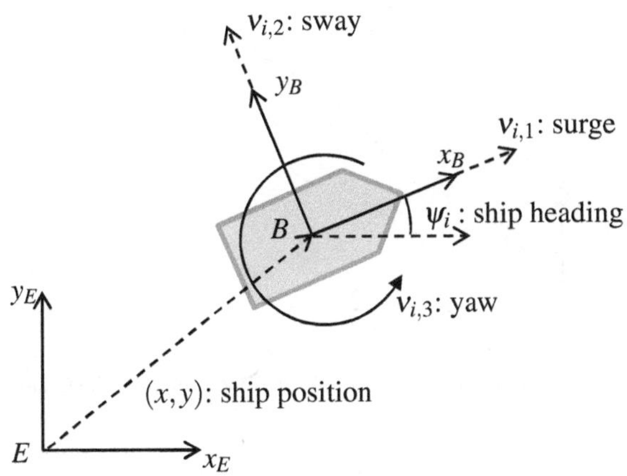

Fig. 7.3 Inertial and body reference frames for a ship.

7.4.2 Trajectory Generation

A group of N agents will have N individual paths where the desired path for agent i is given by $\eta_{di}(\theta_i) = [x_{di}(\theta_i), y_{di}(\theta_i), \psi_{di}(\theta_i)]^T$. To create individual path for each agent, we introduce a Formation Reference Point (FRP) and design a set of designation vectors $l_i \in \mathbb{R}^3$, relative to the FRP, corresponding to the desired formation for agent i, $i = 1, \cdots, N$ (see Fig. 7.4). Let the desired path for the FRP be $\eta_d(\theta) = [x_d(\theta), y_d(\theta), \psi_d(\theta)]^T$. Then agent i will follow the path

$$\eta_{di}(\theta) = \eta_d(\theta) + R(\theta)l_i \tag{7.46}$$

where $R(\theta) = R(\psi_d(\theta))$ denotes the rotation matrix from the frame T to the inertial frame E. In particular, the tangent vector along the path is chosen as the x-axis of the frame T, that is, $x_T = \frac{\partial x_d}{\partial \theta}$ and $y_T = \frac{\partial y_d}{\partial \theta}$. The desired heading $\psi_d(\theta)$ can then calculated as the angle of the tangent vector in the inertial frame

$$\psi_d(\theta) = \arctan\left(\frac{x_T}{y_T}\right) = \arctan\left(\frac{\frac{\partial x_d(\theta)}{\partial \theta}}{\frac{\partial y_d(\theta)}{\partial \theta}}\right). \tag{7.47}$$

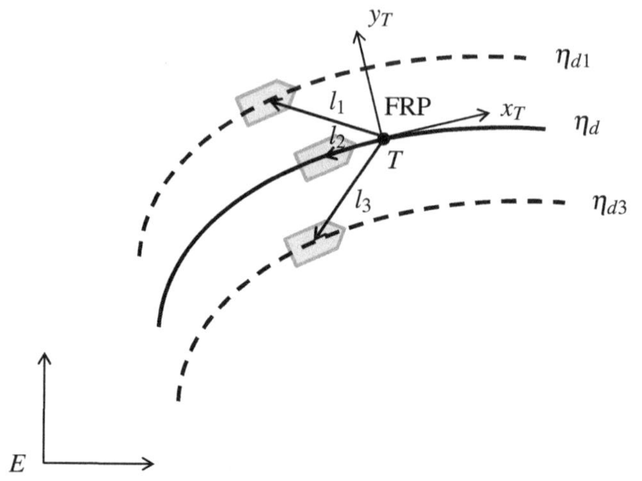

Fig. 7.4 Formation setup using a Formation Reference Point and designation vectors.

An individual agent substitutes its own θ_i into (7.46) and obtains its desired path. When θ_i's are synchronized, these agents are in the desired formation configuration.

7.4.3 Preliminary Backstepping Design

To apply to the results in Section 7.3 to (7.39)-(7.40), we first present a preliminary backstepping design [120] that brings agent dynamics to the form in (7.6). Towards this end, we define the error variables

$$z_{i,1} = \eta_i - \eta_{di}(\theta_i) \tag{7.48}$$
$$z_{2,i} = \nu_i - \alpha_{i,1} \tag{7.49}$$
$$\omega_i = v(t) - \dot{\theta}_i \tag{7.50}$$

where $z_{i,1}$ is the tracking error of agent i and $\alpha_{i,1}$ is the virtual control to be determined later.

Step 1. Differentiating (7.49), we obtain

$$\dot{z}_{i,1} = \dot{\eta}_i - \frac{\partial \eta_{di}}{\partial \theta_i}\dot{\theta}_i \tag{7.51}$$

$$= R_i(\psi_i)(\alpha_i + z_{i,2}) - \frac{\partial \eta_{di}}{\partial \theta_i}\dot{\theta}_i. \tag{7.52}$$

Consider a Lyapunov function

$$V_{i,1} = z_{i,1}^T P_{i,1} z_{i,1}, \tag{7.53}$$

where $P_{i,1} = P_{i,1}^T > 0$. The time derivative of $V_{i,1}$ along the trajectory of (7.52) is given by

$$\dot{V}_{i,1} = 2z_{i,1}^T P_{i,1}\left(R_i(\psi_i)\alpha_i + R_i(\psi_i)z_{i,2} - \frac{\partial \eta_{di}}{\partial \theta_i}\dot{\theta}_i\right). \tag{7.54}$$

We now take the virtual control α_i as

$$\alpha_i = R_i^T(\psi_i)\left(A_{i,1}z_{i,1} + \frac{\partial \eta_{di}}{\partial \theta_i}v(t)\right) \tag{7.55}$$

where $A_{i,1}$ is chosen such that

$$P_{i,1}A_{i,1} + A_{i,1}^T P_{i,1} = -I_3. \tag{7.56}$$

This leads to

$$\dot{V}_{1,i} = -z_{i,1}^T z_{i,1} + 2z_{i,1}^T P_{i,1} R_i(\psi_i) z_{i,2} + 2z_{i,1}^T P_{i,1}\frac{\partial \eta_{di}}{\partial \theta_i}\omega_i. \tag{7.57}$$

Step 2. In this step, we will design τ_i in (7.40). In aid of this, we differentiate α_i in (7.55) as

$$\dot{\alpha}_i = \sigma_i + \frac{\partial \alpha_i}{\partial \theta_i}\dot{\theta}_i \tag{7.58}$$

where σ_i includes all the terms that do not contain $\dot{\theta}_i$:

$$\sigma_i = \dot{R}_i^T(\psi_i)R_i(\psi_i)\alpha_i + R_i^T(\psi_i)\left(A_{i,1}R_i(\psi_i)\nu_i + \frac{\partial \eta_{di}}{\partial \theta_i}\dot{v}(t)\right) \tag{7.59}$$

and

$$\frac{\partial \alpha_i}{\partial \theta_i} = R_i^T(\psi_i)\left(-A_{i,1}\frac{\partial \eta_{di}}{\partial \theta_i} + \frac{\partial^2 \eta_{di}}{\partial \theta_i^2}v(t)\right). \tag{7.60}$$

We next consider the Lyapunov function

$$V_{i,2} = z_{i,2}^T P_{i,2} z_{i,2}, \tag{7.61}$$

where $P_{i,2} = P_{i,2}^T > 0$, and define $V = V_{i,1} + V_{i,2}$. The time derivative of V is then given by

$$\dot{V} = \dot{V}_{i,1} + 2z_{i,2}^T P_{i,2} (\dot{v}_i - \dot{\alpha}_i) . \tag{7.62}$$

From (7.40) , (7.57) and (7.58), we rewrite $\dot{V}$ as

$$\begin{aligned} \dot{V} &= -z_{i,1}^T z_{i,1} + 2z_{i,1}^T P_{i,1} R_i(\psi_i) z_{i,2} + 2z_{i,1}^T P_{i,1} \frac{\partial \eta_{di}}{\partial \theta_i} \omega_i \\ &+ 2z_{i,2}^T P_{i,2} \left(M_i^{-1} (\tau_i - D_i(v_i) v_i) - \sigma_i - \frac{\partial \alpha_i}{\partial \theta_i} \dot{\theta}_i \right) . \end{aligned} \tag{7.63}$$

We then pick

$$\tau_i = D_i(v_i) v_i + M_i \left(A_{i,2} z_{i,2} + \sigma_i + \frac{\partial \alpha_i}{\partial \theta_i} v(t) - P_{i,2}^{-1} R_i^T (\psi_i) P_{i,1} z_{i,1} \right) \tag{7.64}$$

where $A_{i,2}$ is such that

$$P_{i,2} A_{i,2} + A_{i,2}^T P_{i,2} = -I_3 . \tag{7.65}$$

With (7.64), $\dot{V}$ is now given by

$$\dot{V} = -z_{i,1}^T z_{i,1} - z_{i,2}^T z_{i,2} + 2 \left(z_{i,1}^T P_{i,1} \frac{\partial \eta_{di}}{\partial \theta_i} + z_{i,2}^T P_{i,2} \frac{\partial \alpha_i}{\partial \theta_i} \right) \omega_i . \tag{7.66}$$

We further obtain the closed-loop system in the coordinate of $(z_{i,1}, z_{i,2})$ as

$$\begin{pmatrix} \dot{z}_{i,1} \\ \dot{z}_{i,2} \end{pmatrix} = \underbrace{\begin{pmatrix} A_{i,1} & R_i(\psi_i) \\ -P_{i,2}^{-1} R_i^T (\psi_i) P_{i,1} & A_{i,2} \end{pmatrix}}_{F_i} \begin{pmatrix} z_{i,1} \\ z_{i,2} \end{pmatrix} - \underbrace{\begin{pmatrix} -\frac{\partial \eta_{di}}{\partial \theta_i} \\ -\frac{\partial \alpha_i}{\partial \theta_i} \end{pmatrix}}_{g_i} \omega_i \tag{7.67}$$

which is of the form (7.6). Choosing $P = \text{diag}\{P_{i,1}, P_{i,2}\}$, we observe from (7.66) that

$$PF_i + F_i^T P = -I_6 \tag{7.68}$$

which satisfies (7.5). Thus, the preliminary feedback (7.64) brings (7.39)-(7.40) to the form (7.6).

7.4.4 Adaptive Design to Estimate Reference Velocity

We consider a group of four agents following circle-shaped paths. The desired path for agent i is given by

$$\eta_{di}(\theta_i) = \begin{pmatrix} x_{di}(\theta_i) \\ y_{di}(\theta_i) \\ \psi_{di}(\theta_i) \end{pmatrix} = \begin{pmatrix} r_i \cos(\frac{\theta_i}{1200}) \\ r_i \sin(\frac{\theta_i}{1200}) \\ \frac{\theta_i}{1200} + \frac{\pi}{2} \end{pmatrix} \tag{7.69}$$

where r_i is the circling radius of agent i. In the simulation we set $r_1 = 8$, $r_2 = 12$, $r_3 = 16$ and $r_4 = 20$. This implies that after agreement of all path parameters, the vessels will move along different circles parallel to each other.

We assume that only agent 1 has the information of $v(t)$ in (7.6). In the simulation, we choose $v(t) = 10$. According to Remark 7.1, the other agents implement the adaptive design from Chapter 3 to estimate the $v(t)$ information. Since $v(t)$ is constant, this adaptive design takes the following form

$$\dot{\theta}_i = \hat{v}_i - \omega_i \tag{7.70}$$

$$\dot{\hat{v}}_i = -\gamma_i \sum_{i=1}^{\ell} d_{ik} \psi_k(\tilde{\theta}_k), \quad \gamma_i > 0. \tag{7.71}$$

The initial conditions for η_i's are given by $\eta_1(0) = [0\ 0\ \frac{\pi}{2}]^T$, $\eta_2(0) = [1\ -2\ \frac{\pi}{2}]^T$, $\eta_3(0) = [5\ -2\ \frac{\pi}{4}]^T$ and $\eta_4(0) = [0\ -5\ -\frac{\pi}{3}]^T$. The initial velocity for each agent is zero. The communication topology is given by the incidence matrix

$$D = \begin{pmatrix} 0 & 0 & -1 \\ 1 & 0 & 0 \\ -1 & 1 & 0 \\ 0 & -1 & 1 \end{pmatrix} \tag{7.72}$$

which means that only agents 2 and 3, agents 3 and 4, and agent 4 and 1 can exchange their path parameters. We set $\tilde{\theta}_1(0) = 94$, $\tilde{\theta}_2(0) = -137$ and $\tilde{\theta}_3(0) = 90$. The initial conditions for $\hat{v}_i$'s are $\hat{v}_2(0) = -8$, $\hat{v}_3(0) = -10$, $\hat{v}_4(0) = 9$. In (7.9), we choose $\psi_k(x) = x$ and $\mathscr{H}_i$ as a constant gain of 0.5. The control parameter P_i is set to $P_i = \mathrm{diag}\{0.2, 0.2, 1, 10, 10, 40\}$.

Fig. 7.5 shows that before we turn on the adaptation at 100 sec, the agreement of the path variables is not achieved since each agent has different $v(t)$ information. After the adaptation is turned on, the path parameters are synchronized. Five snapshots of the formation are shown in Fig. 7.6. Initially, four agents are inside the smallest circle. Before we turn on the adaptation, the agents are not synchronized on their paths (agents 1 and 4 are ahead of agents 2 and 3). After the adaptation is turned on, the agents achieve the synchronized path following.

7.4.5 Saturation in Thrust

In this example, we assume that the propeller of agent 4 saturates and is only able to produce a surge sped less than the speed assignment. In this case, the saturation constraint of agent 4 will cause steady state errors in the agreement of the path variables. This error is eliminated by employing integral feedback from relative path variables between neighbors. In Design 2, the effect due to the thruster saturation can be reduced with a proportional-integral-derivative (PID) control structure with limited integral and derivative effect, also known as a lead-lag controller with the

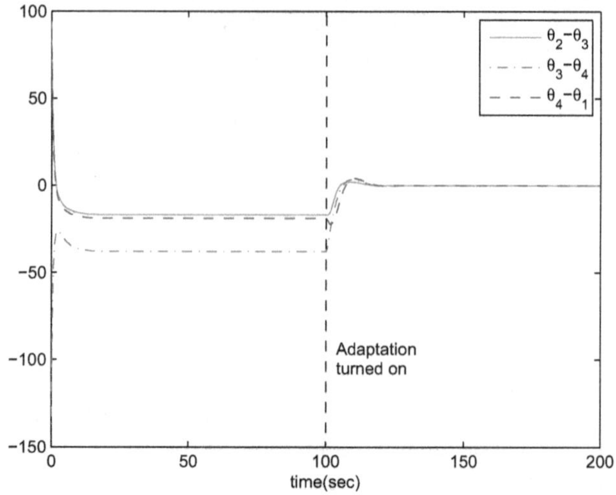

Fig. 7.5 Before adaptation, the path parameters cannot be synchronized since each agent has different $v(t)$ information. After the adaptation is turned on, the synchronization errors converge to zero.

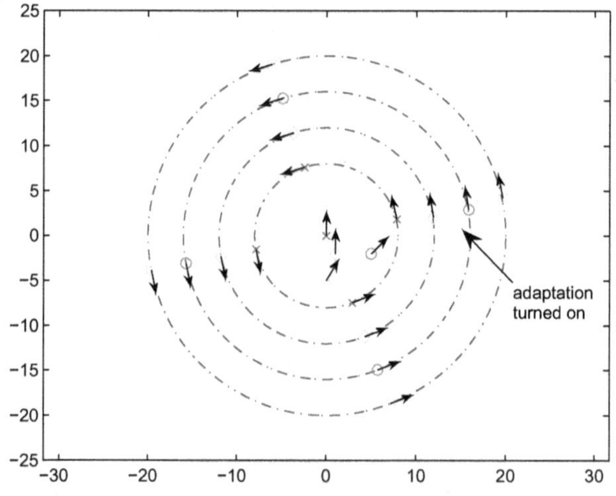

Fig. 7.6 Five snapshots of the formation: Initial formation is inside the smallest circle. Before we turn on the adaptation, the agents are not synchronized on their paths. After adaptation is turned on, synchronized path following is achieved. Agents 1, 2, 3 and 4 are denoted by $\times$, $\diamond$, $\circ$ and $\Box$.

following transfer function:

$$H_{pid,i}(s) = \nu_i \beta_i \frac{1+\mu_i s}{1+\beta_i \mu_i s} \frac{1+T_{d,i}s}{1+\alpha_i T_{d,i}s} \tag{7.73}$$

in which $\nu_i > 0$, $0 \leq T_{d,i} \leq \mu_i$, $1 \leq \beta_i < \infty$, and $0 < \alpha_i \leq 1$. Since (7.73) is Hurwitz and satisfies $Re\left[H_{pid,i}(j\omega)\right] \geq \nu_i > 0$ for all ω, $H_{pid,i}(s)$ falls into the class of input strictly passive systems and stability of the interconnection follows from Theorem 7.2. The control ω_i is then given by

$$\omega_i(s) = H_{pid,i}(s)\tilde{u}_i(s) \tag{7.74}$$

We use Design 2 in Section 7.3.2 and compare the synchronization errors for constant $H_{pid,i}$, i.e., $T_{d,i} = \mu_i = 0$ in (7.73) while $\nu_i = 10$, with the PID structure where $T_{d,i} = 1$, $\mu_i = 2$, $\nu_i = 1$, $\alpha_i = 0.1$ and $\beta_i = 10$. The other parameters are the same as in Section 7.4.4. The synchronization errors $\tilde{\theta}_i$'s shown in Fig. 7.7 illustrate that the PID structure yields a better agreement of the path variables when agent 4 saturates.

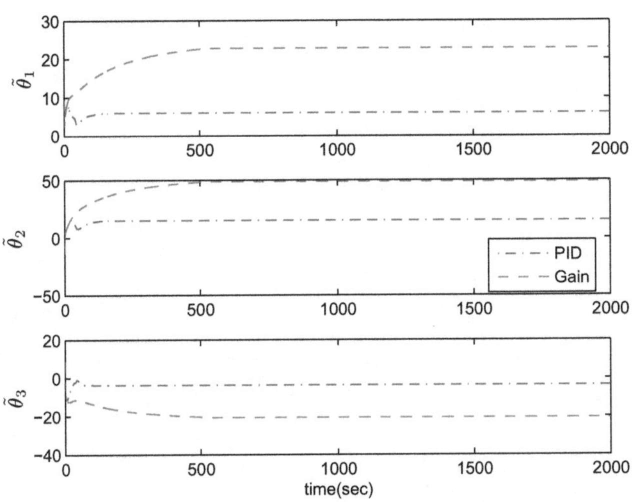

Fig. 7.7 The synchronization errors $\tilde{\theta}_i$ for the PID and the constant gain designs.

7.5 Summary

We applied the passivity-based agreement design in Section 2.5 to a formation tracking scheme where path following systems are synchronized by communicating path variables. The passivation designs offers the flexibility to consider several designs for synchronization where robustness and performance issues can be addressed. The first design only employs the information of the synchronization error and becomes a trajectory tracking control as the path variables reach agreement. The second design makes use of both path error and synchronization error and allows users to prioritize path convergence. We presented a design example that incorporated the basic adaptive design in Chapter 3 and that also explored the flexibility of the passivity-based design to reduce saturation effects.

7.6 Notes

- The path following results in this chapter are mainly based on [59]. Other techniques for path following of a single vehicle include [2, 40, 54, 114, 3, 45].
- The idea of incorporating synchronization/consensus schemes into formation control has also been considered in other studies, e.g., [12, 127, 150].

Chapter 8
Cooperative Load Transport

8.1 Introduction

For the examples presented in previous chapters, the relative information between agents, such as relative distances/positions, relative path variables, is obtained through *explicit* information flow, including sensor measurements and direct communication. While such explicit information flow exists in many cooperative control applications, there are situations where the information flow is *implicit*. For example, suppose that several people move a table and only one person knows where to go. Then, even without explicitly talking to or seeing each other, those people are able to adjust their velocities and forces, and finally succeed in moving the table towards the target. In this example, the communication is implicit, and people receive the information (e.g., where to go, how fast to go) by feeling the contact forces and the trend where the table is going.

In this chapter, we consider a group of agents handling a flexible payload. These agents are modeled as point robots with double integrator dynamics. As the agents move, the payload may be squeezed or stretched, generating contact forces to the agents. The contact forces between the agents and the payload are modeled as gradients of nonlinear potentials that describe the deformations of the payload. Because all the agents are attached to the payload, the contact forces can be considered as *implicit* communication between the agents with the payload acting as the "medium". Our objective is to employ this implicit communication to design decentralized control laws such that the contact forces are regulated at some setpoints and that the agents and the payload move with the same constant velocity in the limit. We assume that the deformations of the payload are so small that the motion of the payload can be approximated as a rigid body. This assumption is reasonable when a rigid load is surrounded with bumpers or elastic materials. Another illustration of this assumption is multiple grippers grasping a rigid load, where the grippers possess compliance from installed flexible mechanisms.

Recall that the formation control designs in Sections 2.6 and 2.7 employ *virtual* attractive/repulsive force feedback between the agents. For our load transport prob-

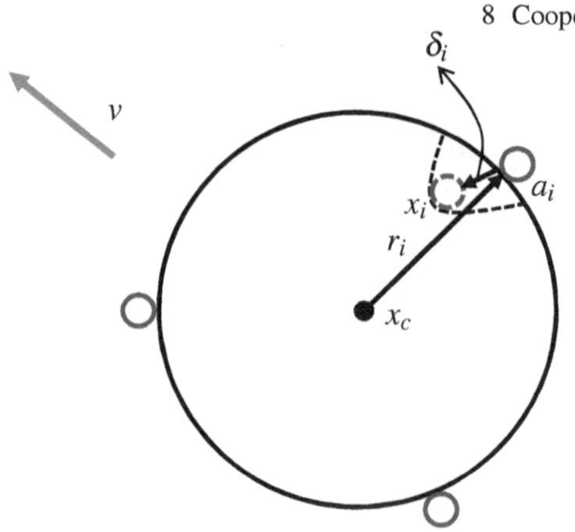

Fig. 8.1 Multiple agents are attached to a common flexible load. Initially the payload is undeformed and agent i is attached to the point a_i. If the load is deformed by agent i, the position of agent i, x_i, is different from a_i and the deformation is approximated by $\delta_i = x_i - a_i$.

lem, the contact forces play the role of *physical* force feedback between the agents and the payload. With this idea in mind, we consider the case where a constant reference velocity is available to all the agents. We propose a decentralized controller that guarantees the force regulation and the velocity convergence of the agents and the payload. This controller, consisting of an internal velocity feedback and an external force feedback from the payload, exhibits a similar structure to the position-based formation control design in Section 2.6. Exploiting this similarity allows us to apply adaptive design techniques to the load transport problem. For example, when no reference velocity is predesigned, we employ the adaptive design results from Chapter 3 and augment the decentralized controller with an integral control term. The resulting control law recovers the traditional integral force control and ensures the agents to achieve the same constant velocity and the force regulation without explicit communication.

8.2 Problem Formulation

Consider N planar agents holding a common flexible load as shown in Fig. 8.1. Each agent is modeled as a point robot. Suppose that the load is initially undeformed and that agent i is attached to the load at the point a_i, i.e., $x_i(0) = a_i(0) = x_c(0) + r_i$, where $x_c \in \mathbb{R}^2$ and $x_i \in \mathbb{R}^2$ are the inertial positions for the center of mass of the load and agent i, and r_i is a fixed vector in the inertial frame. Assuming that the initial orientation of the load θ_c is zero, we define

$$a_i(t) := x_c(t) + R(\theta_c)r_i, \quad R(\theta_c) = \begin{pmatrix} \cos\theta_c & -\sin\theta_c \\ \sin\theta_c & \cos\theta_c \end{pmatrix} \tag{8.1}$$

whose kinematics are given by

$$\dot{a}_i = \dot{x}_c + \dot{\theta}_c \begin{pmatrix} -\sin\theta_c & -\cos\theta_c \\ \cos\theta_c & -\sin\theta_c \end{pmatrix} r_i. \tag{8.2}$$

Note that $a_i(t)$ represents the position where agent i is attached if the payload is undeformed at time t. As the agents move, however, the flexible payload may be squeezed or stretched and therefore $x_i(t) \neq a_i(t)$. The deformation of the payload, approximated by

$$\delta_i = x_i - a_i, \quad i = 1, \cdots, N, \tag{8.3}$$

generates a reaction force f_i to agent i. This reaction force f_i must be zero when the payload is not deformed by agent i, i.e., when $\delta_i = 0$. Define the relative position between agent i and the center of mass of payload as

$$z_i = x_i - x_c, \tag{8.4}$$

and note from (8.1) that

$$\delta_i = z_i - R(\theta_c) r_i. \tag{8.5}$$

We next assume that the reaction force f_i is the gradient of a positive definite and *strictly convex* potential function $\bar{P}_i(z_i)$, that is,

$$f_i = \nabla \bar{P}_i(z_i) \tag{8.6}$$

where $\bar{P}_i(z_i)$ satisfies

$$\bar{P}_i(z_i) = 0 \Longleftrightarrow z_i - R(\theta_c) r_i = 0 \tag{8.7}$$

$$\nabla \bar{P}_i(z_i) = 0 \Longleftrightarrow z_i - R(\theta_c) r_i = 0. \tag{8.8}$$

The strict convexity assumption is satisfied by the linear spring potential model $P_i(\delta_i) = b_i |z_i - R(\theta_c) r_i|^2$, $b_i > 0$, and certain classes of nonlinear models, such as $P_i(\delta_i) = b_i |z_i - R(\theta_c) r_i|^4$.

We further assume that the deformations are small enough so that the dynamics of the payload can be approximated as a rigid body. This assumption is reasonable when a rigid object is surrounded by deformable materials (e.g., bumper) and the agents are attached to those materials. The dynamics of the agents and the payload, restricted to purely translational motion, are given by

$$m_i \ddot{x}_i = F_i - f_i, \quad i = 1, \cdots, N \tag{8.9}$$

$$M_c \ddot{x}_c = \sum_{i=1}^{N} f_i \tag{8.10}$$

where m_i and M_c are the masses of agent i and the load, F_i is the force applied to agent i, and f_i is the contact force defined in (8.6).

The control objective is to design F_i in a decentralized way such that all the agents and the payload converge to the same constant velocity, while the contact forces on the load are regulated, i.e., f_i maintained at a setpoint f_i^d. Because the load must move with a constant speed, f_i^d's are subject to the following constraint:

$$\sum_{i=1}^{N} f_i^d = 0. \tag{8.11}$$

The setpoints f_i^d should also be chosen to satisfy desired properties, such as force closure [94], which ensures that the agents always squeeze the payload at the desired stage. This requires the knowledge of the payload geometry and the grasping points.

In the following sections, we consider the control design in two cases: first, a reference velocity v is predesigned and available to each agent; second, v is not available to each agent. The second case includes as a special case the situation where v is available only to the leader.

8.3 Decentralized Control With Reference Velocity

We note from (8.6) that the reaction force f_i depends on the relative position z_i. If z_i can be regulated to some desired state, f_i would also be maintained accordingly. To this end, we assume that for a given f_i^d, there exists a deformation z_i^d, such that

$$f_i^d = \nabla \bar{P}_i(z_i^d). \tag{8.12}$$

Note that achieving a desired contact force f_i^d is now equivalent to driving the relative position z_i in (8.4) to the desired one z_i^d. Let

$$\xi_c = \dot{x}_c - v, \quad \xi_i = \dot{x}_i - v, \tag{8.13}$$

and

$$\xi = [\xi_1^T, \cdots, \xi_N^T]^T, \quad z = [z_1^T, \cdots, z_N^T]^T, \quad z^d = [(z_1^d)^T, \cdots, (z_N^d)^T]^T. \tag{8.14}$$

Our control objective is thus convergence to the equilibrium $\mathscr{E}_p$

$$\mathscr{E}_p = \left\{ (\xi, \xi_c, z) \mid \xi = 0, \;\; \xi_c = 0, \text{ and } z = z^d \right\}. \tag{8.15}$$

This equilibrium is the same as that of the position-based formation control problem in Section 2.6, where the relative positions between agents are driven to some desired values. Indeed, if the payload is treated as the $N+1$th agent, convergence to (8.15) means that z_i, the relative position between agents i and the payload, is driven to the desired value z_i^d. Thus, the interactions between the $N+1$ agents display a star graph with the payload at the center, as shown in Fig. 8.2. If the contact forces (and thus the relative positions) between agent i, $i = 1, \cdots, N$, and the payload can be regulated, the relative positions between the N agents are maintained tightly.

To achieve the convergence to (8.15), we propose the following control law for agent i

$$F_i = -K_i(\dot{x}_i - v) + f_i^d, \quad K_i = K_i^T > 0. \tag{8.16}$$

This feedback law transforms the agent dynamics (8.9) to the form

$$m_i\ddot{x}_i = -K_i(\dot{x}_i - v) + f_i^d - f_i \tag{8.17}$$

which consists of an internal motion feedback that drives the agent's velocity to v, and an external force feedback that regulates the contact force. We further rewrite (8.17) and (8.10) as

$$\dot{x}_i = \xi_i + v \tag{8.18}$$
$$m_i\dot{\xi}_i = -K_i\xi_i + u_i \tag{8.19}$$
$$\dot{x}_c = \xi_c + v \tag{8.20}$$
$$M_c\dot{\xi}_c = u_c \tag{8.21}$$

where

$$u_i = f_i^d - f_i \tag{8.22}$$

and

$$u_c = -\sum_{i=1}^{N} u_i. \tag{8.23}$$

Note that the agent dynamics (8.18) and (8.19) are of the same form as the position-based formation control in (2.19) and (2.20). In particular, the external feedback u_i in (8.22) is the gradient of the potential function

$$P_i(z_i) = \bar{P}_i(z_i) - \bar{P}_i(z_i^d) - (f_i^d)^T(z_i - z_i^d). \tag{8.24}$$

Thanks to the strict convexity of $\bar{P}_i(z_i)$, $P_i(z_i)$ is positive definite as proven in Lemma 8.1 below and thus plays a similar role to $P_k(z_k)$ in (2.27).

Lemma 8.1. *The potential function (8.24) has a unique global minimum at $z_i = z_i^d$ and is proper.* □

Proof. A direct application of Proposition B.2 in Appendix B.8. □

The payload dynamics (8.20)-(8.21) are almost the same as (8.18)-(8.19), except that (8.21) is only passive from u_c to ξ_c whereas (8.19) is strictly passive from u_i to ξ_i. This is because the payload dynamics (8.10) have no damping. If (8.21) were strictly passive, we could follow the results in Corollary 2.2 and prove the convergence of $\dot{x}_i$ and $\dot{x}_c$ to v and z_i to z_i^d, which means $f_i \to f_i^d$. As demonstrated in the following proposition, these convergence results still remain true when (8.21) is only passive. The proof for this proposition, given in Appendix A.5, relies on the implicit communication topology in Fig. 8.2 and the fact that v is constant.

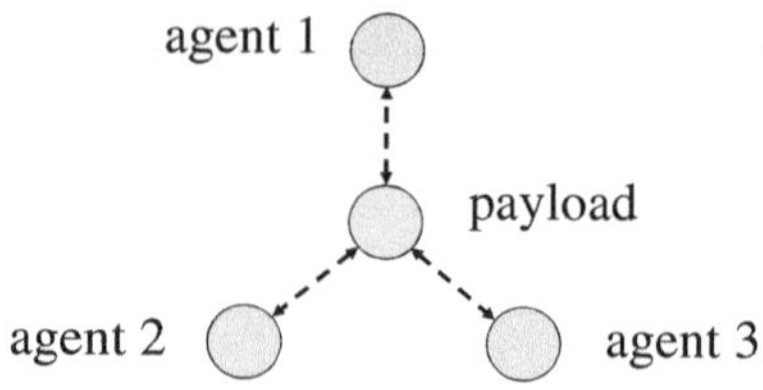

Fig. 8.2 The implicit information topology between three agents and the payload displays a star graph.

Proposition 8.1. *Consider the decentralized control law (8.16) with the dynamics (8.9) and (8.10), where f_i is defined in (8.6). Then, the equilibrium defined in (8.15) is globally asymptotically stable. In particular, $f_i \to f_i^d$, $i = 1, \cdots, N$.* □

8.4 Decentralized Control Without Reference Velocity

In some situations, the reference velocity information may not be available to the agents. For example, the agents may have a common velocity direction but have not agreed on how fast to go. In this case, it is important for the agents to reach a common velocity. To achieve this, we follow the results from Section 3.6 and develop an adaptive control with which each agent estimates the group velocity. We define $\hat{v}_i$ as the velocity estimate for agent i and propose the following update law for $\hat{v}_i$:

$$\dot{\hat{v}}_i = \Lambda_i (f_i^d - f_i) \tag{8.25}$$

in which $\Lambda_i = \Lambda_i^T > 0$. Note that $\hat{v}_i$ stops updating when $f_i^d \equiv f_i$, that is, the contact force is regulated at the desired setpoint. This means that z_i remains constant and thus agent i and the payload have the same velocity. If all the agents have the same velocity as the payload, they move with the same velocity.

We next modify the design in (8.16) as

$$F_i = -K_i(\dot{x}_i - \hat{v}_i) + m_i \dot{\hat{v}}_i + f_i^d. \tag{8.26}$$

With a slight abuse of notation, we let

$$\xi_i = \dot{x}_i - \hat{v}_i \tag{8.27}$$

and transform (8.9) and (8.26) to

$$\dot{x}_i = \xi_i + \hat{v}_i \tag{8.28}$$

$$m_i \dot{\xi}_i = -K_i \xi_i + f_i^d - f_i. \tag{8.29}$$

Define

$$\hat{v} = [(\hat{v}_1)^T, \cdots, (\hat{v}_N)^T]^T, \tag{8.30}$$

and note that the equilibria set of (8.10), (8.28) and (8.29) is given by

$$\mathscr{E}_p^* = \left\{ (\xi, \dot{x}_c, z, \hat{v}) \mid \xi = 0,\ \hat{v} - 1_N \otimes \dot{x}_c = 0 \text{ and } z = z^d \right\}. \tag{8.31}$$

Convergence to $\mathscr{E}_p^*$ means that all the agents maintain the desired contact forces on the payload while reaching the same velocity as the payload. This agreed velocity depends on the initial condition of $\hat{v}$ and the initial velocity of the payload and is characterized in the following proposition:

Proposition 8.2. *Consider the decentralized control laws in (8.25) and (8.26). The trajectories of $(\xi, \dot{x}_c, z, \hat{v})$ remains bounded and converge to $\mathscr{E}_p^*$ in (8.31). In particular, $\dot{x}_c$ and $\dot{x}_i$'s converge to $\bar{v} \in \mathbb{R}^2$, which is the weighted average of the initial payload velocity $\dot{x}_c(0)$ and the initial velocity estimates $\hat{v}_i(0)$, $i = 1, \cdots, N$:*

$$\bar{v} = \left(\sum_{i=1}^{N} \Lambda_i^{-1} + M_c\right)^{-1} \left(M_c \dot{x}_c(0) + \sum_{i=1}^{N} \Lambda_i^{-1} \hat{v}_i(0)\right). \tag{8.32}$$

□

A special example of the design (8.25)-(8.26) is when only one agent, say agent 1, has the v information. In this case, agent 1 can choose to turn off the estimation (8.25) by selecting $\Lambda_1 = 0$ and letting $\hat{v}_1(0) = v$. This leads to the same controller in (8.16) for agent 1. A simple calculation from (8.32) shows $\lim_{\Lambda_1 \to 0} \bar{v} = v$, which means that the other agents asymptotically recover the v information and the group will eventually move with the velocity v.

Proposition 8.3. *Suppose that agent 1 has the v information and implements (8.16) while the other agents apply the control (8.25) and (8.26). Let*

$$\tilde{v}_1 \equiv 0, \quad \tilde{v}_i = \hat{v}_i - v, \quad i = 2, \cdots, N \tag{8.33}$$

and $\tilde{v} = [(\tilde{v}_1)^T, \cdots, (\tilde{v}_N)^T]^T$. Then the equilibrium

$$\mathscr{E}_a^* = \left\{ (\xi, \xi_c, \tilde{v}, z) \mid \xi = 0,\ \xi_c = 0,\ \tilde{v} = 0 \text{ and } z = z^d \right\} \tag{8.34}$$

is globally asymptotically stable. □

The proofs for Propositions 8.2 and 8.3 are given in Appendix A.6 and A.7, respectively.

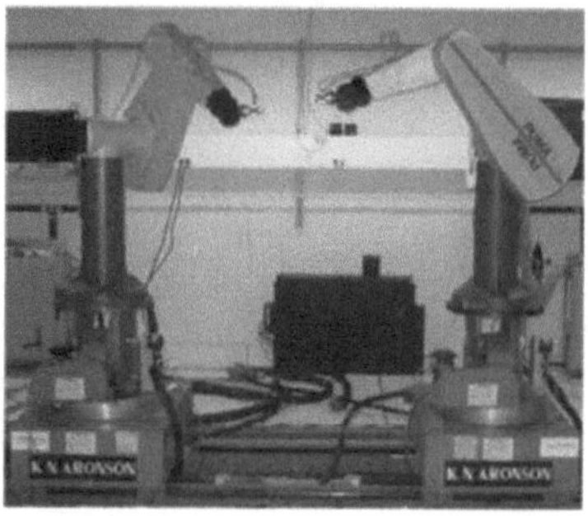

Fig. 8.3 Experiment testbed of two PUMA 560 arms: Mounted on the wrist of each arm is a six-degree freedom force/torque sensor.

8.5 Experiments

8.5.1 Hardware

The experimental testbed, shown in Fig. 8.3, consists of two PUMA 560 arms, each with a six-degree of freedom force/torque sensor mounted on the wrist. Two computers running xPC target perform the real time control of the two arms and the data acquisition is achieved using PCI interface boards. Those two xPC computers run at an update rate of 1 kHz and only do low level tasks. There is a user interface host computer that runs MATLAB and communicates with the control computers through an Ethernet cable using UDP. This configuration allows high level processing and control to be done on the host computer while the control computer implements the low level control loop.

8.5.2 Implementation

Because the end-effector of the PUMA arm is of six-degree freedom, we consider it as a fully actuated agent. To simplify the implementation and reduce the effects due to the uncertainty of the arm inertia, we choose to implement the control laws on the kinematic level rather than the dynamic level. Motivated by a standard singular perturbation analysis of (8.17) for small m_i, we obtain the following controller by setting the right hand side of (8.17) to zero (i.e., setting $\ddot{x}_i = 0$) and solving for $\dot{x}_i$:

$$\dot{x}_i = \Gamma_i(f_i^d - f_i) + v, \quad i = 1,2 \tag{8.35}$$

where $\Gamma_i > 0$, x_1 and x_2 are the positions of the end-effectors of the right and the left PUMA arms. This controller can also be justified by treating the payload as a virtual massless spring and invoking a Lyapunov analysis of (8.35) with the Lyapunov function $\sum_{i=1}^{2} P_i(z_i)$. Likewise, for the adaptive design, we implement

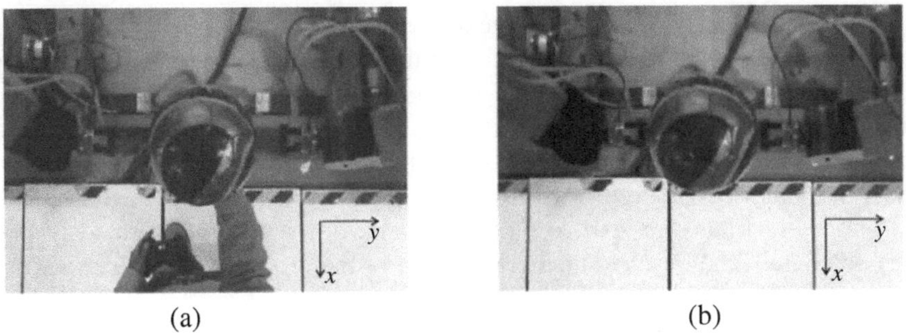

Fig. 8.4 The first two stages of the experiments (pictures were taken from the top view): (a) The two manipulators first move towards the ball. (b) Once the end-effectors reach the ball, we implement the control (8.35) with $v = [0\ 0]^T$ m/s to achieve the static holding. The left (right) PUMA refers to the left (right) manipulator in this figure.

$$\dot{x}_i = \Gamma_i(f_i^d - f_i) + \hat{v}_i \tag{8.36}$$

and keep the update law of $\hat{v}_i$ the same as (8.25).

Once the force measurement f_i is available, $\dot{x}_i$ is computed from (8.35) or (8.36) and transformed to the joint velocities using the pseudoinverse of the manipulator Jacobian matrix, i.e.,

$$\dot{q}_i = J_i^+ \dot{x}_i \tag{8.37}$$

where $\dot{q}_i$ is the joint velocity, and J_i^+ is the pseudoinverse of J_i that satisfies

$$\dot{x}_i = J_i \dot{q}_i. \tag{8.38}$$

The joint positions for the next step are calculated from $\dot{q}_i$ as

$$q_i(k+1) = q_i(k) + T\dot{q}_i \tag{8.39}$$

where $T = 0.001$ sec is the sampling period. The next step joint position is then tracked by a low level PID controller.

Our experiments are performed in the following steps:

Approaching. As seen in Fig. 8.4(a), the two manipulators approach a light weight soccer ball, fed by a person, along the y direction. In this stage, we tune the positions and the orientations of the end-effectors so that they are aligned on the same line. Thus, their motions are restricted to the same plane. We also ensure that the end-effectors are normal to the contact surfaces and that the line connecting both end-effectors approximately passes through the center of the ball. This guarantees no rotational motion when the manipulators squeeze the ball in the next stage.

Static Holding. As the end-effectors reach the ball in Fig. 8.4(b), we turn on the controllers in (8.35) with $v = [0\ 0]^T$ m/s, which implies that the ball will be held statically and squeezed. The squeeze forces are along the y direction and their desired setpoints are chosen to be ± 10 N, that is $f_1^d = [0\ -10]^T$ N and $f_2^d = [0\ 10]^T$

N, thereby satisfying (8.11). These large desired squeeze forces help maintaining the grasp and reducing the rotational motion of the ball. In case that the squeeze force drops below a certain amount (6 N in the experiments), we increase the feedback gain Γ_i to quickly drive f_i back. To ensure that the end-effector does not slip on the contact surface, we discard the force measurements along the x direction, which means that only the squeeze forces are controlled. Once the squeeze forces reach the desired setpoints, we start to move the ball with several basic maneuvers as discussed below and evaluate the performance of the proposed controllers in these cases.

Moving the object. In this step, we examine the proposed controllers in the scenarios of moving with v available, estimation of v, and circular motion. In all cases, we set $\Gamma_i = \Gamma = 0.001$.

8.5.2.1 Moving with v available

After the stage of static holding, we move the object along the x direction by assigning $v = [0.02\ 0]^T$ m/s to each end-effector. The force measurements from the sensors on the wrists are shown in Fig. 8.5. The squeeze forces are maintained at ± 10 N at the stage of static holding (17.5 sec-30 sec) and oscillate more around ± 10 N when the end-effectors start moving (30 sec-47.5 sec). This is partially due to the dynamic effects of the low level PID tracking controller that we ignored in the implementation. The trajectories of both end-effectors from 30 sec to 47.5 sec are shown in Fig. 8.6, where no significant rotational motion is observed. Moreover, we compute from Fig. 8.6 that the approximate average velocities along the x direction is 0.02 m/s, the same as v.

8.5.2.2 Adaptive Estimation of v

We now examine the adaptive design (8.36). We assign $v = -[0\ 0.005]^T$ m/s to the right PUMA while the left PUMA has no v information. Therefore, the left PUMA needs to implement the adaptive design (8.36) to estimate v. The initial estimate of the left PUMA $\hat{v}_1(0)$ is chosen as zero. Since the motion is along y direction, which is the squeeze direction, no rotational motion is generated. The experimental results are shown in Fig. 8.7, where the estimate from the left PUMA converges to -0.005 m/s meanwhile the squeeze forces are well maintained at ± 10 N.

As a comparison, we implement the nonadaptive control law (8.35) with $v = -[0\ 0.005]^T$ m/s for the right PUMA and $v = [0\ 0]^T$ m/s for the left PUMA. Since they do not have the same v information, the experimental results in Fig. 8.8 show the existence of steady state errors in both the reference velocity tracking and the force regulation when the end-effectors are moving (after 30 sec). These results are expected from the analysis in Section 3.2. In fact, the two end-effectors, along the y direction, are governed by

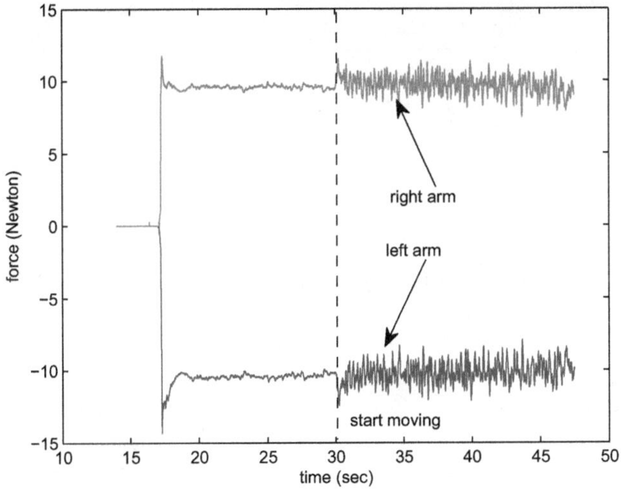

Fig. 8.5 The force measurements along the squeeze direction when the manipulators move in the x direction.

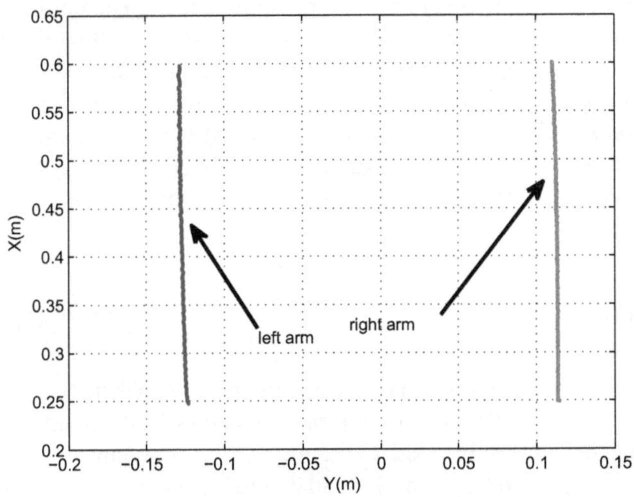

Fig. 8.6 The trajectories for both end-effectors when they are moving forward with $v = [0.02\ 0]^T$ m/s.

$$\dot{x}_{1,y} = \Gamma(f_{1,y}^d - f_{1,y}) - 0.005, \quad \dot{x}_{2,y} = \Gamma(f_{2,y}^d - f_{2,y}) + 0 \tag{8.40}$$

where the subscript y denotes the y component of each vector. Rewriting (8.40) as

$$\dot{x}_{1,y} = \Gamma(\underbrace{f^d_{1,y} - \frac{0.005}{2\Gamma}}_{\bar{f}^d_{1,y}} - f_{1,y}) - 0.0025 \tag{8.41}$$

$$\dot{x}_{2,y} = \Gamma(\underbrace{f^d_{2,y} + \frac{0.005}{2\Gamma}}_{\bar{f}^d_{2,y}} - f_{2,y}) - 0.0025, \tag{8.42}$$

we see that due to the different reference velocities, the desired setpoints $f^d_{i,y}$ are shifted to new setpoints $\bar{f}^d_{i,y}$. It is easy to calculate $\bar{f}^d_{i,y}$ as ± 12.5 N, which matches our results in Fig. 8.8(b).

8.5.2.3 Circular Motion

Although the nonadaptive controller (8.35) is restricted to the case of constant reference velocity, we test it when the end-effectors move in a circular motion. In this case, we choose $f^d_1 = [0 \ -15]^T$ N and $f^d_2 = [0 \ 15]^T$ N and show both the force measurements and the trajectories of the two end-effectors in Fig. 8.9. Fig. 8.9(a) illustrates that once the end-effectors start circling (after 40 sec), the squeeze forces cannot be maintained at ± 15 N due to the periodic motion. The tracking of the reference velocity is not well achieved as the trajectories in Fig. 8.9(b) are not a perfect circle. This is because the force and the motion control are coupled all the time.

To improve the tracking performance, we add deadzone for the force error $f^d_i - f_i$ so that the force and motion control are decoupled when f_i is close to f^d_i. The experimental results in Fig. 8.10 illustrate the improvement of the tracking of the circles at the cost of more fluctuating contact forces in the deadzones.

8.6 Summary

In this chapter, we studied a motion coordination problem where a group of agents move a flexible payload. The contact forces, which describe the relative information between the agents and the payload, build up *implicit* communication in the group. When the desired constant velocity is available to each agent, we developed a decentralized controller that achieves the convergence to the reference velocity and the force regulation. We also considered the situation where the reference velocity is not available. We proposed an adaptive control that recovers the nonadaptive results. Both nonadaptive and adaptive control laws were compared to the formation control designs in Chapters 2 and 3. Experiments were performed and validated our designs.

8.7 Notes

• There exists a considerable amount of literature on the control design for the load transport problem, including the motion/force control [89, 145], event-based control scheme [92], caging without force control [106], leader/follower comply mechanism [154, 57]. Reference [130] employed screw theory to examine a system of two non-holonomic wheeled mobile manipulators holding a common load. In [129] multi-gripper grasping was considered, where one of the grippers is rigid and the others are flexible with built-in linear springs, and stabilization control laws were developed to achieve both position and force control. The authors of [128] considered transporting large (possibly flexible) objects as an impedance control problem and performed experiments using multiple mobile robots with manipulators to validate their controllers. The work in [134] studied modeling and manipulating totally deformable objects and the solution is centralized and based on finite element model. Our model of the payload is different from [134] since we assume the payload to be a partially flexible object that deforms only around the grasping points while [134] considered the deformations of all particles on the flexible object.

• In contrast with the existing literature, we solved the load transport problem in a similar way to the position-based formation control problem. The goal of this chapter is to address in the same passivity-based framework the connection between the formation control problem, where the interaction force is virtual, with the multi-agent load carrying problem, where the interaction force is physical.

• Expanding the dynamics (8.9) with the adaptive control law (8.25) and (8.26), we obtain

$$m_i\ddot{x}_i = -K_i(\dot{x}_i - \hat{v}_d^i(0)) + m_i\dot{\hat{v}}_i + f_i^d - f_i + \underbrace{K_i\Lambda_i \int (f_i^d - f_i)}_{\text{integral force control}} \tag{8.43}$$

which is of the integral force control form [89, 145]. Such an integral force control has been shown in [145] to be robust with respect to small time delay in the force measurements.

• We have considered only translational motion of the agents and the payload in this chapter. This is a simplifying assumption, which allows us to illustrate in a basic form the connection between the formation control problem and the load transport problem. Indeed, if the grasp is rigid, the agents are capable to exert torques to the payload. Then the dynamics for the orientation of the payload become

$$I_c\ddot{\theta}_c = \sum_{i=1}^{N} r_{ic}{}^{\times} f_i + \sum_{i=1}^{N} \tau_i \tag{8.44}$$

where I_c is the inertia of the payload, τ_i is the torque transmitted to the payload from agent i, $r_{ic} = x_i - x_c$, and

$$r_{ic}{}^{\times} f_i = r_{ic}^x f_i^y - r_{ic}^y f_i^x, \tag{8.45}$$

in which $r_{ic} = [r_{ic}^x, r_{ic}^y]^T$ and $f_i = [f_i^x, f_i^y]^T$. Our assumption on pure translational motion means that τ_i's may be chosen to stabilize θ_c to a constant. The design of τ_i may require the information of r_{ic} and θ_c. Once θ_c is stabilized, the remaining motion would be only translational. Therefore, the formulation in this chapter only reflects the translational part of the load transport problem.

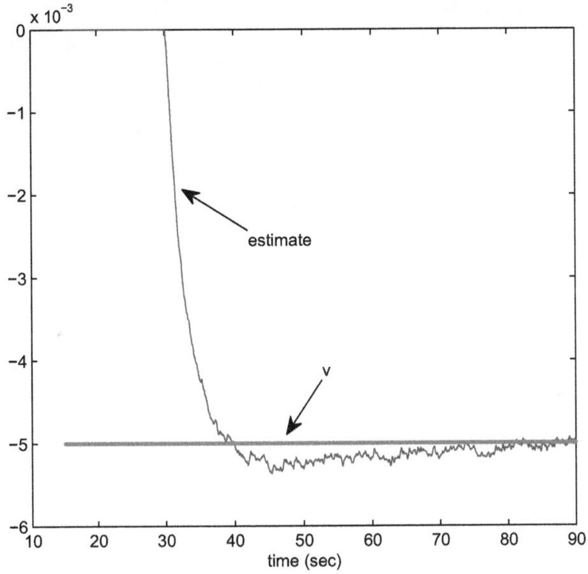

(a)

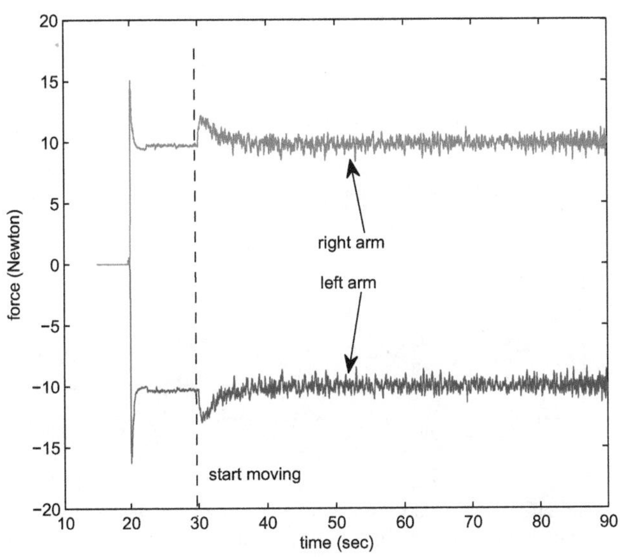

(b)

Fig. 8.7 Estimation of v: The estimate of v converges to its true value in Fig. 8.7(a) while the squeeze force measurements are well-maintained at ± 10 N as in Fig. 8.7(b).

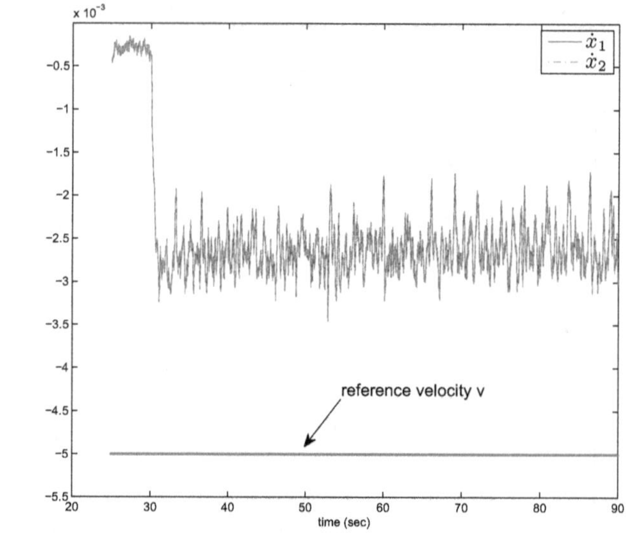

(a)

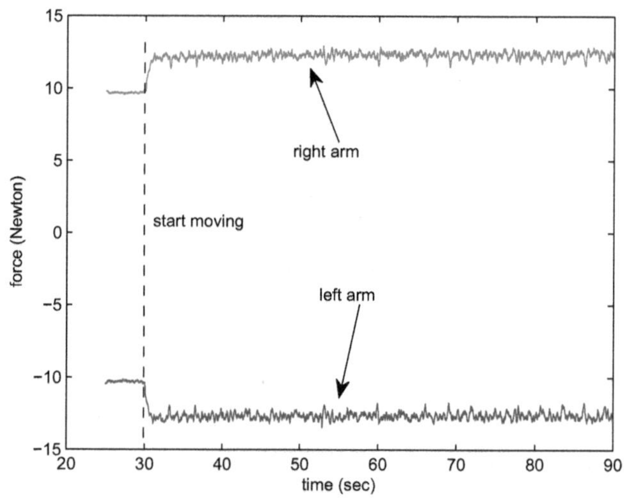

(b)

Fig. 8.8 When the end-effectors have different v information, the tracking of the reference velocity is not achieved as in Fig. 8.8(a) and the contact forces cannot be maintained at the desired ones as in Fig. 8.8(b).

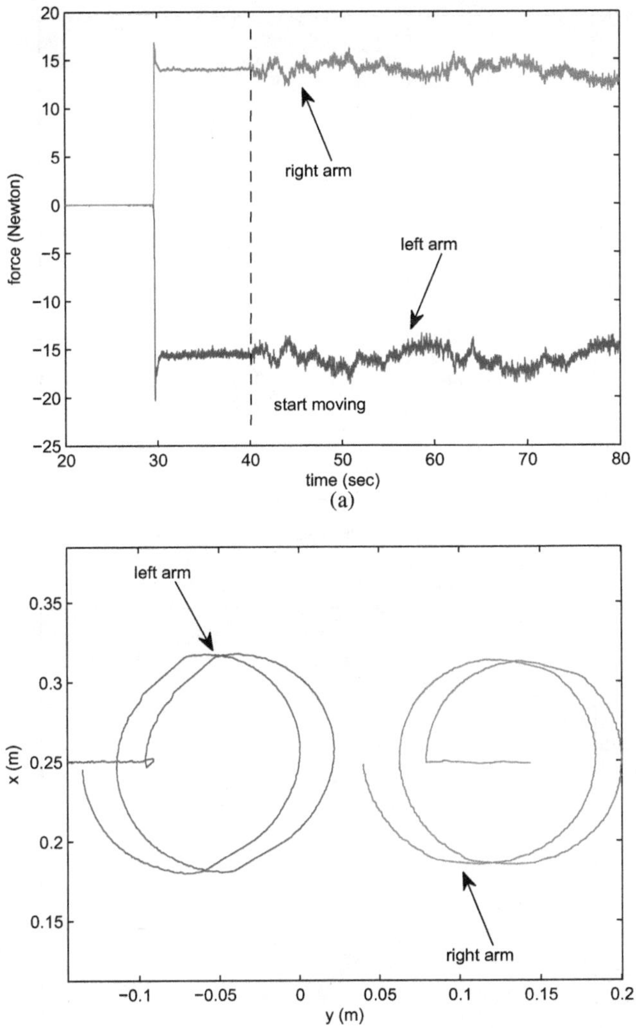

(b)

Fig. 8.9 The squeeze forces cannot be maintained at the desired value due to the circular motion of the end-effectors. The trajectories of the end-effectors in Fig. 8.9(b) show that the tracking of the reference velocity is not well achieved.

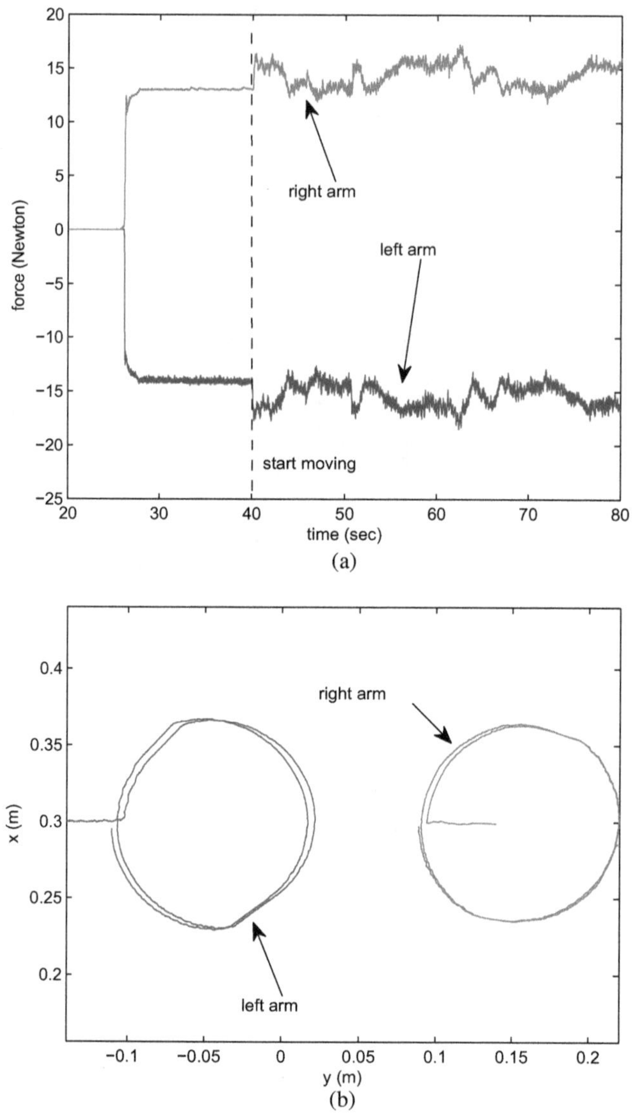

Fig. 8.10 With deadzones added to the force control, the tracking of the reference velocity is improved as in Fig. 8.10(b) while the squeeze forces are more fluctuating in Fig. 8.10(a).

Chapter 9
Caveats for Robustness

9.1 Introduction

In this chapter we consider the position-based formation control design in (2.74) and investigate its robustness with respect to switching topology, link gain variation and unmodeled dynamics. For convenience, we rewrite (2.74) here

$$(M \otimes I_p)\ddot{\mathbf{x}} + (K \otimes I_p)\dot{\mathbf{x}} + (L_\Delta \otimes I_p)\mathbf{x} = 0 \tag{9.1}$$

where $M = \text{diag}\{m_1, \cdots, m_N\}$, $K = \text{diag}\{k_1, \cdots, k_N\}$ and $L_\Delta = D\Delta D^T$ is the weighted Laplacian. Recall that (9.1) ensures global asymptotic stability of the origin of $\dot{\mathbf{x}}$ and $\mathbf{z} = (D^T \otimes I_p)\mathbf{x}$.

We first analyze (9.1) with switching topologies. Such switching may occur due to the vehicles joining or leaving a formation, transmitter/receiver failures, limited communication/sensor range, or physical obstacles temporarily blocking sensing between vehicles. For single integrator dynamics, switching topology has been studied in [63, 103] and stability under arbitrary switching has been ascertained for classes of coordination algorithms. In contrast, for second order dynamics, we illustrate with an example that a destabilizing switching sequence that triggers instability exists. We then show that stability is maintained when switching is sufficiently fast or slow so that is does not interfere with the natural frequencies of the group dynamics.

We next investigate stability properties when the link weights are perturbed by small sinusoidal oscillations. To illustrate this instability in its most basic form, we make a simplifying assumption that the perturbation is sinusoidal and transform the group dynamics into a form that reveals a parametric resonance mechanism [52, 96, 53]. This transformation employs the spectral properties of the graph Laplacian and decouples the relative motion from the motion of the center of the agents. When mass inertia and damping terms are identical for all agents, we obtain decoupled *Mathieu equations* [96], which make parametric resonance explicit. For broader classes of mass and damping matrices, we obtain coupled Mathieu equations and discuss which frequencies lead to parametric resonance. Next, we show that sinusoidal perturbations do not destabilize the system if they are slow or fast enough. The sinusoidal perturbations studied in this situation are not necessarily the most commonly occurring ones in practice. However, they allow us to study worst-case scenarios to deepen the understanding of fundamental stability and robustness properties in cooperative systems.

We finally study the effect of input unmodeled dynamics, such as fast actuator dynamics. Following standard singular perturbation arguments, we prove that the stability of the nominal design that ignores the effects of unmodeled dynamics is preserved when the stable unmodeled dynamics are sufficiently fast. As we illustrate with an example, how fast the unmodeled dynamics must be is dictated by the graph structure and the mass inertia matrix.

9.2 Instability due to Switching Topology

9.2.1 Example

Consider four agents with an undirected graph that switches between a ring graph and a complete graph[1]. Let $M = I$, $K = kI$ and $\Delta = \delta I$ for some constants $k > 0$ and $\delta > 0$. Then, the closed-loop dynamics (9.1) become

$$\ddot{\mathbf{x}} + k\dot{\mathbf{x}} + \delta(L_i \otimes I_p)\mathbf{x} = 0 \quad i = 1,2 \tag{9.2}$$

where $L_i = D_i D_i^T$ is the Laplacian matrix for the ring graph when $i = 1$, and for the complete graph when $i = 2$.

Because L_1 and L_2 admit the same set of orthonormal eigenvectors q_j, $j = 1, \cdots, 4$ for their eigenvalues $\{0,2,2,4\}$ and $\{0,4,4,4\}$, respectively, the change of variables $d_j = (q_j^T \otimes I_p)\mathbf{x}$, $j = 1, \cdots, 4$ decouples the dynamics (9.2) into

$$\ddot{d}_j + k\dot{d}_j + \delta\lambda_{j_i} d_j = 0, \tag{9.3}$$

where λ_{j_i} is the jth eigenvalue of the Laplacian L_i, $i = 1,2$. It then follows from standard results in switching systems [81, 1, 80] that, if the damping k is small, and if $\delta\lambda_{j_1} < 1$ and $\delta\lambda_{j_2} > 1$, then (9.3) is destabilized by a switching sequence that selects $i = 1$ when $d_j^T \dot{d}_j > 0$ and $i = 2$ otherwise. Instability with this sequence follows from the Lyapunov-like function $V = ||d_j||^2 + ||\dot{d}_j||^2$ which increases along the trajectories of (9.3). Because the eigenvalues λ_{2_i} and λ_{3_i} switch between the values 2 and 4 in our example, if $\delta \in (1/4, 1/2)$, then $\delta\lambda_{j_1} < 1$ and $\delta\lambda_{j_2} > 1$ indeed hold for $j = 2,3$. This means that, when the damping is small, a destabilizing switching sequence exists.

We demonstrate this instability with a simulation in Fig. 9.1. We choose $p = 1$ and four agents. Although the system (9.1) guarantees agreement of $\mathbf{x}_i$'s for any

[1] A complete graph is a graph where every two nodes are connected.

fixed connected graph, when the communication topology switches between a complete graph and a ring graph according to the sequence described above, Fig. 9.1 shows that the relative distances between the agents diverge.

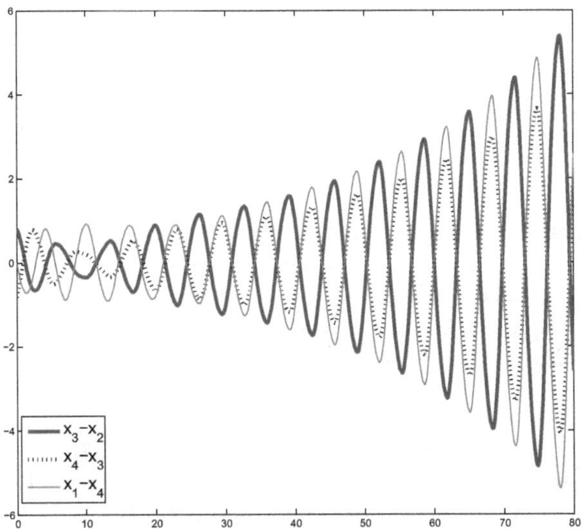

Fig. 9.1 A switching sequence described in Section 9.2.1 between the ring and complete graphs destabilizes the relative positions between the agents in the system (9.1).

9.2.2 Comparison with First-order Agent Models

The instability example presented in the previous section occurs only when the agent dynamics are second or higher order. In this section, we show that for agents modeled as first order integrators, switching between connected graphs will not lead to instability. In fact, the agreement of $\mathbf{x}_i$'s can be achieved even if the graph loses connectivity pointwise in time. Note that for first order agents, the $\mathscr{H}_i$'s in Fig. 2.2 are simply static passive blocks. We then restrict our attention to the following class of first order agreement protocols

$$\dot{\mathbf{x}}_i = -\sum_{k=1}^{M} d_{ik}(t)\,\psi_k(\mathbf{z}_k) \quad \mathbf{z}_k := \sum_{j=1}^{N} d_{jk}(t)\mathbf{x}_j, \tag{9.4}$$

rewritten in vector form (2.3), (2.24) as

$$\dot{\mathbf{x}} = -(D(t) \otimes I_p)\, \psi(\mathbf{z}) \tag{9.5}$$
$$\mathbf{z} = (D(t)^T \otimes I_p)\mathbf{x} \tag{9.6}$$

where the multivariable nonlinearity $\psi(\cdot)$ has the property (2.46). The matrix $D(t)$ is piecewise continuous because its entries exhibit step changes when a change occurs in the communication topology. We define the time varying graph Laplacian as

$$L(t) = D(t)D(t)^T. \tag{9.7}$$

If the graph remains connected for all $t \geq 0$, that is, if

$$\lambda_2\{L(t)\} \geq \sigma_c > 0 \quad \forall t \geq 0 \tag{9.8}$$

for some constant $\sigma_c > 0$ that does not depend on time, then it is not difficult to show that $\mathbf{x}_i$'s in (9.5)-(9.6) reach an agreement despite the time-varying $L(t)$. We now prove agreement under a less restrictive *persistency of excitation* condition which stipulates that graph connectivity be established over a period of time, rather than pointwise in time:

Proposition 9.1. *Consider the system (9.5)-(9.6) where $\mathbf{x} \in \mathbb{R}^{pN}$ comprises of the components $\mathbf{x}_i \in \mathbb{R}^p$, $i = 1, \cdots, N$ concatenated as in (2.3), $\psi(\cdot)$ satisfies (2.46), and $D(t)$ is piecewise continuous incidence matrix. Let S be an $(N-1) \times N$ matrix with orthonormal rows that are each orthogonal to 1_N; that is,*

$$S 1_N = 0 \quad SS^T = I_{N-1}. \tag{9.9}$$

If there exist constants $\delta > 0$ and $\alpha > 0$ such that, for all $t_0 \geq 0$,

$$\int_{t_0}^{t_0+\delta} SL(t)S^T dt \geq \alpha I, \tag{9.10}$$

where $L(t)$ is defined in (9.7), then the protocol (9.5)-(9.6) achieves the agreement of $\mathbf{x}_i$'s. □

The proof of this proposition can be found in Appendix A.8. The "persistency of excitation" condition (9.10) means that $SL(t)S^T$ is nonsingular when integrated over a period of time, and not necessarily pointwise in time. Since, by construction of S in (9.9), $SL(t)S^T$ inherits all eigenvalues of $L(t)$ except the one at zero, its smallest eigenvalue is

$$\lambda_1\{SL(t)S^T\} = \lambda_2\{L(t)\}, \tag{9.11}$$

which means that nonsingularity of the matrix $SL(t)S^T$ is equivalent to connectivity of the graph. Because Proposition 9.1 does not require nonsingularity of $SL(t)S^T$ pointwise in time, it allows the graph to lose pointwise connectivity as long as it is established in the integral sense of (9.10). The pointwise connectivity situation (9.8) is a special case of Proposition 9.1 because, then, (9.10) readily holds with $\alpha = \sigma_c \delta$.

9.2.3 When is Stability Maintained?

Having demonstrated stability robustness of first order agreement protocol with respect to switching topology, we now come back to the second order protocol (9.1) and consider under what conditions the stability of (9.1) is maintained.

Since (9.1) guarantees the origin of $(\dot{\mathbf{x}},\mathbf{z})$ to be exponentially stable for a fixed connected graph, using the concept of dwell-time [81, 91, 56], we can ensure $\dot{\mathbf{x}} \to 0$ and $\mathbf{z} \to 0$ if all graphs in the switching sequence are connected and if the interval between consecutive switchings is no shorter than some minimum dwell time $\tau > 0$, where estimates for τ can be obtained following [56]. We next employ the concept of an "average graph" to show that fast and periodic switching also preserves stability.

Consider a periodic switching sequence $\sigma(t)$ in which the topology switches $n-1$ times, $n \geq 1$, during one period T. We label n graph Laplacians in T as L^i_Δ, $i = 1,\cdots,n$ and denote their dwell times by τ_i, $i = 1,\cdots,n$, $\sum_{i=1}^n \tau_i = T$. We thus study the switched system:

$$(M \otimes I_p)\ddot{\mathbf{x}} + (K \otimes I_p)\dot{\mathbf{x}} + (L_\Delta^{\sigma(t)} \otimes I_p)\mathbf{x} = 0 \tag{9.12}$$

where

$$L_\Delta^{\sigma(t)} \in \{L_\Delta^1, L_\Delta^2, \cdots, L_\Delta^n\}. \tag{9.13}$$

To determine the stability of (9.12)-(9.13), we investigate the eigenvalues of the state transition matrix evaluated over a period T:

$$\Xi(T,0) = e^{A_N\tau_N} \cdots e^{A_2\tau_2} e^{A_1\tau_1}, \tag{9.14}$$

where

$$A_i = \begin{pmatrix} \mathbf{0}_N & I_N \\ -M^{-1}L_\Delta^i & -M^{-1}K \end{pmatrix} \otimes I_p \tag{9.15}$$

is the system matrix of (9.12) in the coordinates of $(\mathbf{x},\dot{\mathbf{x}})$, $i = 1,\cdots,N$. When τ_i's are small, we rewrite (9.14) as

$$\begin{aligned}
\Xi(T,0) &= \prod_{i=1}^{n}[I + \tau_i A_i + O(\tau_i^2)] \\
&= I + \sum_{i=1}^{N} \tau_i A_i + O(T^2) \\
&= I + TA^{av} + O(T^2)
\end{aligned} \tag{9.16}$$

where

$$A^{av} = \begin{pmatrix} \mathbf{0}_N & I_N \\ -M^{-1}L_\Delta^{av} & -M^{-1}K \end{pmatrix} \otimes I_p \tag{9.17}$$

and

$$L_\Delta^{av} = \frac{1}{T}\sum_{i=1}^{n} \tau_i L_\Delta^i \tag{9.18}$$

is the average of the n graph Laplacians during the period T.

Because the linear combination (9.18) preserves the structure of a Laplacian, L_Δ^{av} defines an average graph obtained by superimposing the individual graphs $i = 1,\cdots,n$. In this average graph, the links are weighted by τ_i/T, which represents the relative dwell time of each graph constituting the average. This means that, if the time-varying graph is *jointly connected* as in [63], then the averaged graph described by L_Δ^{av} is connected. We point out that the connectedness of L_Δ^{av} also satisfies the persistency of excitation condition in (9.10) with $\delta = T$ since for all $t_0 \geq 0$

$$\int_{t_0}^{t_0+T} SL_\Delta^{\sigma(t)} S^T dt = SL_\Delta^{av} S^T, \tag{9.19}$$

which is positive definite if and only if the average graph is connected.

We finally show that, when T is sufficient small, connectedness of the average graph implies stability of (9.12)-(9.13). To see this, note from (9.16) that the eigenvalues of $\Xi(T,0)$ are given by

$$\kappa_i = 1 + T\lambda_i + O(T^2), \quad i = 1,\cdots,2N, \tag{9.20}$$

where λ_i's are the eigenvalues of A^{av}. It follows that if the graph induced by the averaged Laplacian L_Δ^{av} is connected, then all λ_i's have negative real parts, except the one, say λ_1, at zero. This zero eigenvalue results from the null space of A^{av}, spanned by $a = [1_N^T \quad 0_N^T]^T$, which is also the null space of A_i, $i = 1,\cdots,n$. We thus conclude that $\Xi(T,0)a = a$, which implies $\kappa_1 = 1$. Then, for sufficiently small T, κ_i in (9.20), $i = 2,\cdots,2N$, remain inside the unit circle and $\kappa_1 = 1$ corresponds to the motion of the center, thereby guaranteeing the asymptotic stability of the subspace spanned by $a = [1_N^T \quad 0_N^T]^T$. Note that convergence to this subspace guarantees $\dot{\mathbf{x}} \to 0$ and $\mathbf{z} \to 0$.

Lemma 9.1. *Consider the closed loop dynamics (9.12)-(9.13) with a switching signal $\sigma(t)$ of period T. If the averaged graph induced by (9.18) is connected, then there exists a T^*, such that for $T < T^*$, the the subspace spanned by $a = [1_N^T \quad 0_N^T]^T$ is asymptotically stable.* □

9.3 Parametric Resonance

9.3.1 Example

To illustrate parametric resonance in its most basic form, we study an example of the cooperative system (9.1) with $M = I$, $K = kI$ and $\Delta = \delta I$. To further simplify the notation we consider the single degree-of-freedom case $p = 1$. The same analysis extends to $p > 1$ with the use of Kronecker algebra. The graph is now time-invariant but the link gain δ is perturbed by a cosine term $\varepsilon \cos \omega t$, thus leading to the closed-loop model

$$\ddot{\mathbf{x}} + k\dot{\mathbf{x}} + (\delta + \varepsilon \cos \omega t)L\mathbf{x} = 0. \tag{9.21}$$

Note from Property 1.2 that L can be diagonalized by an orthonormal matrix Q:

$$Q^T LQ = L_d := \text{diag}\{\lambda_N, \cdots, \lambda_1\} \tag{9.22}$$

where $\lambda_N \geq \lambda_{N-1} \geq \cdots \geq \lambda_1 = 0$. If follows from Property 1.3 that if the graph is connected, then only λ_1 is zero and the corresponding column in Q is $\frac{1}{\sqrt{N}}1_N$ due to Property 1.1. Thus, we let

$$Q = [S^T \ \frac{1}{\sqrt{N}}1_N] \tag{9.23}$$

where S satisfies (9.9), and decompose $\mathbf{x}$ as

$$\mathbf{x} = S^T d + \frac{1_N}{\sqrt{N}}c, \tag{9.24}$$

where $d \in \mathbb{R}^{N-1}$ and $c \in \mathbb{R}$.

The dynamics of c correspond to the evolution of the center of $\mathbf{x}$ and is obtained by premultiplying (9.21) by $\frac{1}{\sqrt{N}}1_N^T$:

$$\ddot{c} + k\dot{c} = 0. \tag{9.25}$$

The solution $c(t)$ approaches $\dot{c}(0)/k + c(0)$, which means that the time-varying link gains do not affect the motion of the center.

Next we derive the dynamic equations for d. Since $SS^T = I_{N-1}$, we obtain from (9.24) that

$$d = S\mathbf{x} \tag{9.26}$$

which, from (9.21), results in

$$\ddot{d} + k\dot{d} + (\delta + \varepsilon \cos \omega t)SL\mathbf{x} = 0. \tag{9.27}$$

We further note from (9.24) that

$$SL\mathbf{x} = SLS^T d \tag{9.28}$$

and from (9.22)-(9.23) that

$$SLS^T = \bar{L}_d \tag{9.29}$$

where $\bar{L}_d = \text{diag}\{\lambda_N, \cdots, \lambda_2\}$. Substituting (9.28)-(9.29) into (9.27), we obtain

$$\ddot{d}_j + k\dot{d}_j + (\delta + \varepsilon \cos \omega t)\lambda_{n+1-j}d_j = 0, \quad j = 1, \cdots, N-1, \tag{9.30}$$

which is a Mathieu equation [52, 149, 96] with the natural frequency $\sqrt{\delta\lambda_{N+1-j}}$. It then follows from standard results for the Mathieu equation that instability occurs when the frequency of the perturbation is around $\omega = 2\sqrt{\delta\lambda_i}/r$, $r = 1, 2, 3, \cdots$, for each $i = 2, \cdots, N$. When damping k is zero, parametric resonance occurs at these frequencies for arbitrarily small ε. For nonzero damping k, parametric resonance occurs for sufficiently large values of ε.

9.3.2 Coupled Mathieu Equations

In the previous example, the assumptions that $M = I$ and $K = kI$ played a crucial role in obtaining the decoupled Mathieu equations (9.30). We now remove this assumption and study the case where M, K and Δ in (9.1) are diagonal matrices with not necessarily identical entries. We then reveal parametric resonance with an analysis of *coupled* Mathieu equations as in [96, Section 5.4], [149, 53, 52]. When each link gain δ_i is perturbed by $\varepsilon\bar{\delta}_i \cos\omega t$, (9.1) becomes

$$M\ddot{\mathbf{x}} + K\dot{\mathbf{x}} + D(\Delta + \varepsilon\cos\omega t\bar{\Delta})D^T\mathbf{x} = 0 \tag{9.31}$$

where $\bar{\Delta} = \mathrm{diag}\{\bar{\delta}_1, \cdots, \bar{\delta}_\ell\}$. Premultiplying by the inverse of M, we obtain

$$\ddot{\mathbf{x}} + M^{-1}K\dot{\mathbf{x}} + M^{-1}L_\Delta\mathbf{x} + \varepsilon\cos\omega t M^{-1}L_{\bar{\Delta}}\mathbf{x} = 0. \tag{9.32}$$

where $L_{\bar{\Delta}} = D\bar{\Delta}D^T$. The coordinate transformation $y = \mathscr{T}^{-1}\mathbf{x}$, where $\mathscr{T}$ is composed of the eigenvectors of $M^{-1}L_\Delta$, then leads to

$$\ddot{y} + \mathscr{T}^{-1}M^{-1}K\mathscr{T}\dot{y} + \Lambda y + \varepsilon\cos\omega t\,\mathscr{T}^{-1}M^{-1}L_{\bar{\Delta}}\mathscr{T}y = 0, \tag{9.33}$$

in which

$$\Lambda = \mathrm{diag}\{\hat{\lambda}_N, \cdots, \hat{\lambda}_1\} \tag{9.34}$$

and $\hat{\lambda}_i$'s are the eigenvalues of $M^{-1}L_\Delta$. Because a similarity transformation brings $M^{-1}L_\Delta$ to the symmetric form $M^{-\frac{1}{2}}L_\Delta M^{-\frac{1}{2}}$, we conclude that $\hat{\lambda}_i$'s are real and nonnegative. Because $\mathscr{N}(D^T)$ is spanned by 1_N, one of the eigenvalues of $M^{-1}D\Delta D^T$, say $\hat{\lambda}_1$, is zero and the corresponding column in $\mathscr{T}$ is 1_N. Similarly to (9.23)-(9.24), we rewrite $\mathscr{T}$ as

$$\mathscr{T} = [S \quad 1_N] \tag{9.35}$$

and note that

$$\mathbf{x} = \mathscr{T}y = Sd + 1_N c \tag{9.36}$$

where $d \in \mathbb{R}^{n-1}$, and $c \in \mathbb{R}$ is the center of $\mathbf{x}$. It then follows from (9.33) and the decomposition (9.36) that

$$\ddot{y} + \mathscr{T}^{-1}M^{-1}K\mathscr{T}\dot{y} + \Lambda y + \varepsilon\cos\omega t\,\mathscr{T}^{-1}M^{-1}L_{\bar{\Delta}}Sd = 0, \tag{9.37}$$

since $1_N c$ lies in $\mathscr{N}(D^T)$.

When the damping term K is small, the off-diagonal entries of $\mathscr{T}^{-1}M^{-1}K\mathscr{T}$ can be neglected [30], that is,

$$\mathscr{T}^{-1}M^{-1}K\mathscr{T} \approx \mathrm{diag}\{\bar{k}_1, \cdots, \bar{k}_N\} := \bar{K} \tag{9.38}$$

where $\bar{k}_i$ is the ith diagonal entry of $\mathscr{T}^{-1}M^{-1}K\mathscr{T}$. The dynamics in (9.37) can then be written as

$$\begin{pmatrix} \ddot{d} \\ \ddot{c} \end{pmatrix} = -\bar{K}\begin{pmatrix} \dot{d} \\ \dot{c} \end{pmatrix} - \Lambda\begin{pmatrix} d \\ c \end{pmatrix} - \varepsilon\cos\omega t\begin{pmatrix} S^*M^{-1}L_{\bar{\Delta}}S & 0 \\ \zeta M^{-1}L_{\bar{\Delta}}S & 0 \end{pmatrix}\begin{pmatrix} d \\ c \end{pmatrix} \tag{9.39}$$

where $\mathscr{T}^{-1} = \begin{pmatrix} S^* \\ \zeta \end{pmatrix}$.

We note from (9.39) that the dynamics of d are decoupled from that of c and that stability of the relative motion of the agents is determined by the d-dynamics. Results for coupled Mathieu equations in [149, 96, 52] applied to (9.39) indicate that parametric resonance occurs around the frequencies

$$\omega = \frac{\sqrt{\hat{\lambda}_j} \pm \sqrt{\hat{\lambda}_k}}{r} \quad j \neq k, \quad j,k = 2,\cdots,N. \tag{9.40}$$

and

$$\omega = \frac{2\sqrt{\hat{\lambda}_j}}{r}, \quad j = 2,\cdots,N, \quad r = 1,2,3\cdots \tag{9.41}$$

For $\bar{K} \neq 0$, parametric resonance occurs at these frequencies if ε is sufficiently large. The parametric resonance resulting from (9.40) is known as *Combination Resonance* because the excitation frequency ω is a linear combination of two natural frequencies $\sqrt{\hat{\lambda}_j}$ and $\sqrt{\hat{\lambda}_k}$ [149]. When (9.41) is satisfied, the corresponding mode, d_{N-j+1}, is excited and the resulting parametric resonance is called *Subharmonic Resonance*. Such resonances are well studied in structural mechanics literature and are not further discussed here.

9.3.3 Fast Varying Perturbation

In the examples above instability occurs when the frequency of the perturbation interferes with the natural frequencies of the cooperative system. We now show that if the perturbation is fast enough (i.e., large ω), the origin of $(\dot{\mathbf{x}}, \mathbf{z})$ is asymptotically stable. In the next subsection, we investigate slow perturbations.

Defining $\tau_f = \omega t$ and denoting

$$\frac{d(\cdot)}{d\tau_f} = (\cdot)', \tag{9.42}$$

we rewrite the perturbed model in (9.32) as

$$\omega^2\mathbf{x}'' + \omega M^{-1}K\mathbf{x}' + M^{-1}(L_\Delta + \varepsilon\cos\tau_f L_{\bar{\Delta}})\mathbf{x} = 0. \tag{9.43}$$

Using the new variables $z_f = \mathbf{z}(\tau)/\omega$, and $v_f = \mathbf{x}'$, we obtain from (9.43) that

$$\begin{pmatrix} v_f' \\ z_f' \end{pmatrix} = \frac{1}{\omega} \underbrace{\begin{pmatrix} -M^{-1}K & -M^{-1}D(\Delta + \varepsilon \cos \tau_f \bar{\Delta}) \\ D^T & \mathbf{0}_\ell \end{pmatrix}}_{A^f(\tau_f)} \begin{pmatrix} v_f \\ z_f \end{pmatrix}. \tag{9.44}$$

When ω is sufficiently large, the averaging method [69] is applicable to (9.44) and the average of $A^f(\tau_f)$ is given by

$$A^f_{av} = \frac{1}{2\pi} \int_0^{2\pi} A^f(t)dt \tag{9.45}$$

$$= \begin{pmatrix} -M^{-1}K & -M^{-1}D\Delta \\ D^T & \mathbf{0}_\ell \end{pmatrix}, \tag{9.46}$$

which is the system matrix of (9.1) written in the coordinate of $(\dot{\mathbf{x}}, \mathbf{z})$. Therefore, A^f_{av} is asymptotically stable. The following lemma is thus a consequence of Theorem B.9 in Appendix B.9:

Lemma 9.2. *Consider the closed-loop system (9.31). There exists a $\omega_f > 0$ such that for $\omega > \omega_f$, the origin of $(\dot{\mathbf{x}}, \mathbf{z})$ is asymptotically stable.* □

9.3.4 Slowly Varying Perturbation

To analyze the system (9.32) with slowly varying perturbation (small ω), we look at its system matrix $A_s(t)$ in the $(\dot{\mathbf{x}}, \mathbf{z})$-coordinates:

$$A_s(t) = \begin{pmatrix} -M^{-1}K & -M^{-1}D(\Delta + \varepsilon \cos \omega t \bar{\Delta}) \\ D^T & \mathbf{0}_\ell \end{pmatrix}. \tag{9.47}$$

Note that $(\dot{\mathbf{x}}, \mathbf{z})$ is restricted to the following subspace

$$S_x = \{(\dot{\mathbf{x}}, \mathbf{z}) | \dot{\mathbf{x}} \in \mathbb{R}^{Np}, \mathbf{z} \in \mathscr{R}(D^T \otimes I_p)\}. \tag{9.48}$$

For any fixed t, if $\Delta + \varepsilon \cos \omega t \bar{\Delta} > \mathbf{0}_\ell$, that is

$$0 \le \varepsilon < \min_{i=1,\cdots,\ell} \frac{\delta_i}{\bar{\delta}_i}, \tag{9.49}$$

it follows that the origin of $(\dot{\mathbf{x}}, \mathbf{z})$ is asymptotically stable on S_x, which implies that $A_s(t)$ restricted to S_x is Hurwitz.

We next evaluate the derivative of $A_s(t)$ as

$$\dot{A}_s(t) = \begin{pmatrix} \mathbf{0}_N & \varepsilon \omega \sin \omega t M^{-1} D \bar{\Delta} \\ \mathbf{0}_{\ell \times N} & \mathbf{0}_\ell \end{pmatrix} \tag{9.50}$$

and compute its norm:

$$\|\dot{A}_s\| = \varepsilon\omega|\sin(\omega t)|\sqrt{\lambda_{\max}\begin{pmatrix} \mathbf{0}_N & \mathbf{0}_{N\times\ell} \\ \mathbf{0}_{\ell\times N} & \Delta D^T M^{-2} D\Delta \end{pmatrix}} \tag{9.51}$$

$$= \varepsilon\omega|\sin(\omega t)|\sqrt{\lambda_{\max}(\Delta D^T M^{-2} D\Delta)} \tag{9.52}$$

$$\leq \varepsilon\omega\sqrt{\lambda_{\max}(\Delta D^T M^{-2} D\Delta)}. \tag{9.53}$$

Since $\|\dot{A}\|$ is bounded, we conclude from Theorem B.10 in Appendix B.10 that for sufficiently small ω or ε, the origin of $(\dot{\mathbf{x}}, \mathbf{z})$ of the perturbed system (9.32) is asymptotically stable.

Lemma 9.3. *Consider the closed-loop system (9.31). There exists a $\bar{\mu} > 0$ such that for $\varepsilon\omega < \bar{\mu}$, the origin of $(\dot{\mathbf{x}}, \mathbf{z})$ is asymptotically stable.* □

9.4 Unmodeled Dynamics

We consider the following closed-loop system with unmodeled dynamics, $i = 1, \cdots, N$,

$$m_i\ddot{\mathbf{x}}_i = C_i\xi_i \tag{9.54}$$

$$\varepsilon\dot{\xi}_i = A_i\xi_i + B_i\tau_i \tag{9.55}$$

where (9.55) represents the unmodeled dynamics, $\varepsilon > 0$, A_i is Hurwitz, and τ_i is defined as

$$\tau_i = -k_i\dot{\mathbf{x}}_i - \sum_{j=1}^{\ell} d_{ij}\delta_j\mathbf{z}_j. \tag{9.56}$$

When ε is small, the unmodeled dynamics are fast. We further assume that the dc gain of the unmodeled dynamics is $C_iA_i^{-1}B_i = -I$ so that the reduced model obtained by setting $\varepsilon = 0$ in (9.54)-(9.55) is identical to (9.1). It then follows from standard singular perturbation arguments (see [69, Example 11.14] reviewed in Appendix B.11) that there exists ε^* such that for $\varepsilon < \varepsilon^*$, the origin of $(\dot{\mathbf{x}}, \mathbf{z})$ is asymptotically stable.

To illustrate the dependence of ε^* on the graph and the mass inertia, we simplify the model in (9.54)-(9.55) by assuming $M^{-1}K = kI_p$, $\Delta = \delta I_\ell$, $A = -I_p$, $B = I_p$ and $C = I_p$:

$$m_i\ddot{\mathbf{x}}_i = \xi_i \tag{9.57}$$

$$\varepsilon\dot{\xi}_i = -\xi_i + \tau_i. \tag{9.58}$$

Denoting $\bar{\xi} = (M^{-1} \otimes I_p)\xi$, we rewrite (9.57)-(9.58) in the compact form:

$$\begin{pmatrix} \dot{\mathbf{x}} \\ \ddot{\mathbf{x}} \\ \dot{\xi} \end{pmatrix} = \left(\underbrace{\begin{pmatrix} \mathbf{0}_N & I_N & \mathbf{0}_N \\ \mathbf{0}_N & \mathbf{0}_N & I_N \\ -\frac{\delta}{\varepsilon}(M^{-1}L) & -\frac{k}{\varepsilon}I_N & -\frac{1}{\varepsilon}I_N \end{pmatrix}}_{A} \otimes I_p \right) \begin{pmatrix} \mathbf{x} \\ \dot{\mathbf{x}} \\ \xi \end{pmatrix}. \tag{9.59}$$

Then, it is not difficult to show that the $3N$ eigenvalues of A are the roots of the following N characteristic polynomials:

$$s^3 + \frac{1}{\varepsilon}s^2 + \frac{k}{\varepsilon}s + \frac{\delta}{\varepsilon}\bar{\lambda}_i = 0, \quad i = 1, \cdots, N, \tag{9.60}$$

where $\bar{\lambda}_i$'s are the eigenvalues of $M^{-1}L$. A Routh-Hurwitz argument further shows that the exact stability region in the parameter space is given by

$$\varepsilon < \varepsilon^* = \frac{k}{\delta \bar{\lambda}_{\max}}, \tag{9.61}$$

where $\bar{\lambda}_{\max}$ is the maximal eigenvalue of $M^{-1}L$. For sufficiently small ε, (9.61) is satisfied and guarantees stability despite the unmodeled dynamics. Denoting $m_{\min} = \min_i m_i$, we note that a conservative upper bound of $\bar{\lambda}_{\max}$ is $\frac{N}{m_{\min}}$, which implies from (9.61) that if $\varepsilon < \frac{k m_{\min}}{\delta N}$, the origin of $(\dot{\mathbf{x}}, \mathbf{z})$ is stable.

Note that, since $\bar{\lambda}_{\max}$ is the maximal eigenvalue of $M^{-1}L$, ε^* depends not only on the graph structure, but also on the mass distribution of the agents. To illustrate this dependence, we consider four agents with $k = 2$, $\delta = 1$ and $p = 1$. We compare ε^*'s under two graphs as in Fig. 9.2. When $M = \mathrm{diag}\{5, 3, 2, 1\}$, we compute from (9.61) $\varepsilon^* = 1.4797$ for the star graph and $\varepsilon^* = 0.8154$ for the string graph, which means that the star graph is more robust for this M. However, when $M = I_4$, $\varepsilon^* = 0.5, 0.5858$, respectively, for the star graph and the string graph, which implies that the star graph is now less robust.

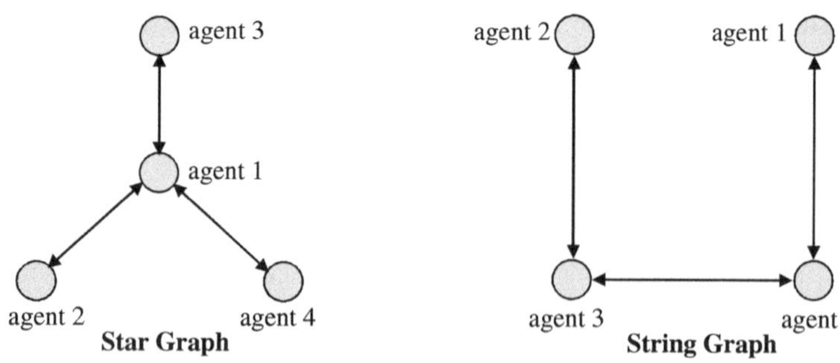

Fig. 9.2 The two graphs used in Section 9.4 to illustrate the dependence of ε^* on the graph structure and mass distribution.

9.5 Summary

In this chapter, we investigated robustness of (2.74) with respect to switching topology, link gain variation and unmodeled dynamics. We illustrated with an example that switching topology can lead to instability and showed that the closed-loop stability is maintained when switching is sufficiently fast and periodic. As a comparison, we also demonstrated that first order agreement protocols have the stability robustness with respect to switching topology. We next revealed a parametric resonance mechanism by transforming the cooperative system with time-varying link gains into Mathieu equations. As in the case of switching graphs, stability is maintained when the sinusoidal perturbation is slow or fast enough that it does not interfere with the natural frequencies of the group dynamics. We finally showed that for fast stable input unmodeled dynamics, the stability of the nominal design is preserved.

Robustness of cooperative control protocols is an area that requires further investigation. Besides the three instability mechanisms presented in this chapter, other instability mechanisms should be revealed and robust redesigns need to be developed.

Appendix A
Proofs

A.1 Proof of Corollary 3.2

We consider V_f and V_b in (2.35) and the following storage function

$$V_p(\varpi) := \frac{1}{2}\varpi^T \varpi. \tag{A.1}$$

We obtain from (2.6) and (3.72)

$$\dot{z} = (D^T \otimes I_p)(y + \hat{v}) \tag{A.2}$$

where

$$\hat{v} = [\hat{v}_1^T, \hat{v}_2^T, \cdots, \hat{v}_N^T]^T. \tag{A.3}$$

We compute $\dot{V}_f$ as

$$\begin{aligned} \dot{V}_f &= \psi^T (D^T \otimes I_p)(y + \hat{v}) \\ &= \{(D \otimes I_p)\psi\}^T (y + \hat{v}) \\ &= -u^T (y + \hat{v}). \end{aligned} \tag{A.4}$$

The time derivative of V_b is given in (2.40). Using Lemma 3.1, we obtain

$$\dot{V}_p = u^T \hat{v}. \tag{A.5}$$

Thus, the time derivative of $V = V_f + V_b + V_p$ yields

$$\dot{V} = -\sum_{i \in \mathscr{I}} W_i(\xi_i) - \sum_{i \notin \mathscr{I}} u_i^T y_i \leq 0 \tag{A.6}$$

which proves the boundedness of $(z(t), \xi(t), \varpi(t))$. We next apply the Invariance Principle and analyze the largest invariant set where $\dot{V} = 0$. Applying the same analysis as in the proof of Theorem 3.1, we conclude from $\dot{V} = 0$ that $\xi = 0$ and

$u=0$, which proves the convergence of $(z(t),\xi(t))$ to $\mathscr{E}$, where $\mathscr{E}$ is as in (2.33). In particular, when Property 2.1 holds, $u=0$ means $z_k \in \mathscr{A}_k$ and thus all trajectories $(z(t),\xi(t))$ starting in $\mathscr{G}$ converge to the set $\mathscr{A}$.

If condition 1 or 2 in Corollary 3.1 holds, $u=0$ implies $\dot{z}=(D^T \otimes I_p)\dot{x}=0$. From (3.72) and the fact that $\mathscr{N}(D^T)$ is spanned by 1_N, we further obtain $|\hat{v}_i(t)-\bar{v}(t)| \to 0$ for some $\bar{v}(t)$.

If in (3.29)-(3.30), $B_i = B$, we note that

$$\sum_{i=1}^{N} \hat{v}_i(t) = B \sum_{i=1}^{N} \varpi_i \tag{A.7}$$

whose time derivative is

$$\sum_{i=1}^{N} \dot{\hat{v}}_i(t) = B \sum_{i=1}^{N} \dot{\varpi}_i = B(A \sum_{i=1}^{N} \varpi_i + B^T \sum_{i=1}^{N} u_i). \tag{A.8}$$

It follows from $u=-(D\otimes I_p)\psi$ and $D^T 1_N = 0$ that $\sum_{i=1}^N u_i = (1_N \otimes I_p)^T u = 0$. Thus, letting $\chi = \frac{1}{N}\sum_{i=1}^N \varpi_i$, we obtain (3.73) from (3.29). Since $|\hat{v}_i - \bar{v}(t)| \to 0$, $\forall i$, (3.74) follows from (A.7).

A.2 Proof of Corollary 3.3

The proof is similar to Appendix A.1. Using the same storage functions as (2.35) and (A.1), we obtain

$$\dot{V}_f = -(u + L^v(y+\hat{v}))^T (y+\hat{v}), \tag{A.9}$$

$\dot{V}_b$ as in (2.40) and $\dot{V}_p$ as in (A.5). It follows that

$$\dot{V}_f + \dot{V}_b + \dot{V}_p = -\sum_{i\in\mathscr{I}} W_i(\xi_i) - \sum_{i\notin\mathscr{I}} u_i^T y_i - (y+\hat{v})^T L^v_{\text{sym}}(y+\hat{v}) \le 0. \tag{A.10}$$

where we recall that $L^v_{\text{sym}} = \frac{1}{2}(L^v + (L^v)^T)$ is positive semidefinite and has the only null space spanned by 1_N if G^v is strongly connected and balanced.

Using the Invariance Principle, we further prove $\xi \to 0$, $y \to 0$, $u \to 0$ and $|\hat{v}_i - \hat{v}_j| \to 0$. The rest of the proof mimics the proof in Appendix A.1.

A.3 Proof of Lemma 4.2

To prove Lemma 4.2 we use standard tools in multivariable calculus, and the results in unconstrained optimization theory. We let γ_1 and γ_2 be the Lipschitz constants of $\nabla F(x)$ and $\nabla^2 F(x)$, respectively, for $x \in \mathscr{D} \subseteq \mathbb{R}^2$. Then, the finite difference approximations in (4.49) and (4.50) satisfy [35, Ch.5]

$$||G_k - \nabla F(x_k)||_1 \leq \frac{1}{2}\gamma_1 h_k \tag{A.11}$$

$$||H_k - \nabla^2 F(x_k)||_1 \leq \frac{5}{3} n\gamma_2 h_k. \tag{A.12}$$

We prove Lemma 4.2 by induction. For $k = 0$, we first show that H_0 is nonsingular. To this end, we let $\varepsilon \leq r$, $h_k \leq \bar{h}$ and

$$\varepsilon + \frac{5}{3} n\bar{h} \leq \frac{\nu}{\beta\gamma_2}, \quad \nu < 1, \tag{A.13}$$

where $\nu < 1$ is to be selected. To show that H_0 is nonsingular, we employ the Banach Lemma given in [67, Theorem 1.2.1]. We first show that, with ε and $\bar{h}$ satisfying (A.13), $||\nabla^2 F(x^*)^{-1} H_0 - I|| < 1$ as follows:

$$\begin{aligned}
||\nabla^2 F(x^*)^{-1} H_0 - I|| &= ||\nabla^2 F(x^*)^{-1}(H_0 - \nabla^2 F(x^*))|| \\
&\leq ||\nabla^2 F(x^*)^{-1}|| \left[||H_0 - \nabla^2 F(x_0)|| + ||\nabla^2 F(x_0) - \nabla^2 F(x^*)||\right] \\
&\leq \beta \left(\frac{5}{3} n\gamma_2 \bar{h} + \gamma_2 \varepsilon\right) \\
&\leq \beta\gamma_2 \left(\varepsilon + \frac{5}{3} n\bar{h}\right) \leq \nu < 1.
\end{aligned} \tag{A.14}$$

Hence, the Banach Lemma implies

$$H_0^{-1} \leq \frac{||\nabla^2 F(x^*)^{-1}||}{1 - ||\nabla^2 F(x^*)^{-1} H_0 - I||} \leq \frac{\beta}{1-\nu}. \tag{A.15}$$

Thus, H_0 is nonsingular, and x_1 is well defined. Next we derive a bound on $|x_1 - x^*|$ as

$$\begin{aligned}
|x_1 - x^*| &= |x_0 + H_0^{-1} G_0 - x^*| \\
&\leq ||H_0^{-1}|| \Big\{ |H_0(x_0 - x^*) - \nabla^2 F(x_0)(x_0 - x^*)| \\
&\quad + |\nabla F(x_0) - \nabla F(x^*) + \nabla^2 F(x_0)(x_0 - x^*)| \\
&\quad + |G_0 - \nabla F(x_0)| \Big\} \\
&\leq \frac{\beta}{1-\nu} \left(\tfrac{5}{3} n\gamma_2 \bar{h} |x_0 - x^*| + \tfrac{\gamma_2}{2} |x_0 - x^*|^2 + \tfrac{\gamma_1 \bar{h}}{2} \right) \\
&\leq \frac{\beta\gamma_2}{1-\nu} \left(\frac{5}{3} n\bar{h} + \frac{\varepsilon}{2}\right) |x_0 - x^*| + \frac{\beta\gamma_1 \bar{h}}{2(1-\nu)} \\
&\leq \frac{\nu}{1-\nu} |x_0 - x^*| + \frac{\beta\gamma_1 \bar{h}}{2(1-\nu)},
\end{aligned} \tag{A.16}$$

which further restricts ν to be $\nu < \frac{1}{2}$ for convergence. Note that, we want $|x_1 - x^*| \leq \varepsilon$ as well, hence ν, $\bar{h}$ and ε must satisfy:

$$\frac{\nu}{1-\nu}\varepsilon + \frac{\beta\gamma_1\bar{h}}{2(1-\nu)} \leq \varepsilon \Longrightarrow \varepsilon(1-2\nu) \geq \beta\gamma_1\bar{h}. \tag{A.17}$$

We thus pick $\bar{h}$ as

$$\bar{h} \leq \frac{2\varepsilon(1-2\nu)}{\beta\gamma_1}, \tag{A.18}$$

which is feasible when $\nu < \frac{1}{2}$. Then, it follows from (A.13) that ν must satisfy

$$\begin{aligned} \varepsilon + \frac{5}{3}n\bar{h} &= \varepsilon + \frac{\frac{10}{3}n\varepsilon(1-2\nu)}{\beta\gamma_1} \leq \frac{\nu}{\beta\gamma_2} \\ &\Longrightarrow \nu \geq \frac{\varepsilon(\frac{10}{3}n\gamma_2 + \beta\gamma_1\gamma_2)}{\gamma_1 + \frac{20}{3}n\varepsilon\gamma_2}, \end{aligned} \tag{A.19}$$

which is consistent with the condition $\nu < \frac{1}{2}$ for a sufficiently small ε. If $\bar{h}$, ε, and ν are selected such that (A.13), (A.18), (A.19) are satisfied, then

$$|x_1 - x^*| \leq \varepsilon \Longrightarrow x_1 \in \mathbf{B}(x^*, \varepsilon)$$

and (A.14) and (A.15) hold for H_1 as well. Induction on $k+1$ goes similarly, and yields

$$|x_{k+1} - x^*| \leq c^{k+1}|x_0 - x^*| + \bar{h}\delta \sum_{n=0}^{k} c^n \tag{A.20}$$

where $c := \frac{\nu}{1-\nu}$ and $\delta := \frac{\beta\gamma_1}{2(1-\nu)}$. Because $\nu < \frac{1}{2}$ implies $c < 1$, the sum in (A.20) converges to $\sum_{n=0}^{\infty} c^n = \frac{1}{1-c}$ and $\lim_{k\to\infty} |x_k - x^*| = O(\bar{h})$.

A.4 Proof of Theorem 5.2

We denote by $\tilde{\beta}_i$ the error variable

$$\tilde{\beta}_i = \bar{\beta}_i - \beta \quad i = 2, \cdots, N \tag{A.21}$$

where $\beta = [(\beta^1)^T, \cdots, (\beta^r)^T]^T$ and note from (5.55) that

$$\dot{\tilde{\beta}}_i = \Lambda_i(\Phi(t) \otimes I_3)u_i \quad i = 2, \cdots, N. \tag{A.22}$$

For consistency, we set $\tilde{\beta}_1 \equiv 0$.

We then rewrite (5.10), (5.53) and (5.54) as

$$^1\mathscr{J}_1 \Delta\dot{\omega}_1 + \Delta\omega_1 \times {}^1\mathscr{J}_1\, {}^1\omega_1 = -f_1\Delta\omega_1 + u_1 \tag{A.23}$$

$$^i\mathscr{J}_i \Delta\dot{\bar{\omega}}_i + \Delta\bar{\omega}_i \times {}^i\mathscr{J}_i\, {}^i\omega_i = -f_i\Delta\bar{\omega}_i + u_i \quad i = 2, \cdots, N. \tag{A.24}$$

To prove the stability of the closed loop system described by the adaptive scheme (A.23), (A.24), (5.18) and (5.55), we investigate the passivity properties of the interconnected system. To this end, we take the Lyapunov function

$$V_a = (q_0 - 1_\ell)^T (q_0 - 1_\ell) + q_v^T q_v + \frac{1}{2}\Delta\omega_1^{T\,1}\mathcal{J}_1\Delta\omega_1 + \frac{1}{2}\sum_{i=2}^{N}(\Delta\bar{\omega}_i)^{T\,i}\mathcal{J}_i\Delta\bar{\omega}_i + \frac{1}{2}\sum_{i=2}^{N}\tilde{\beta}_i^T\Lambda_i^{-1}\tilde{\beta}_i. \tag{A.25}$$

The derivative of V_a along (5.7), (5.55), (A.23) and (A.24) yields

$$\dot{V}_a = -\sum_{i=1}^{N}{}^i\omega_i^T u_i + \Delta\omega_1^T u_1 - f_1\|\Delta\omega_1\|^2 + \sum_{i=2}^{N}\{\Delta\bar{\omega}_i^T u_i - f_i\|\Delta\bar{\omega}_i\|^2\} + \sum_{i=2}^{N}\tilde{\beta}_i^T(\Phi(t)\otimes I_3)u_i \tag{A.26}$$

We further note from (5.51) and (5.52) that

$$(\Phi(t)^T\otimes I_3)\tilde{\beta}_i = {}^i\bar{\omega}_i^d - \omega^d \tag{A.27}$$

from which we rewrite (A.26) as

$$\begin{aligned}\dot{V}_a &= -\sum_{i=1}^{N}{}^i\omega_i^T u_i + \sum_{i=1}^{N}\Delta\omega_i^T u_i - f_1\|\Delta\omega_1\|^2 - \sum_{i=2}^{N}f_i\|\Delta\bar{\omega}_i\|^2\\ &= -\omega^d(t)^T\sum_{i=1}^{N}u_i - f_1\|\Delta\omega_1\|^2 - \sum_{i=2}^{N}f_i\|\Delta\bar{\omega}_i\|^2\\ &= -f_1\|\Delta\omega_1\|^2 - \sum_{i=2}^{N}f_i\|\Delta\bar{\omega}_i\|^2 \le 0,\end{aligned} \tag{A.28}$$

where we used the fact $\sum_{i=1}^N u_i = 0$. The negative semidefinite derivative $\dot{V}_a$ implies that the trajectories $\{q_0, q_v, \Delta\omega_1, \Delta\bar{\omega}_i, \tilde{\beta}_i\}$, $i=2,\cdots,N$ are bounded. We further conclude that $\Delta\omega_1 \to 0$ and $\Delta\bar{\omega}_i \to 0$, $i=2,\cdots,N$ from Theorem B.5. By taking the derivative of (A.23) and (A.24), we note that $\Delta\ddot{\omega}_1$ and $\Delta\ddot{\bar{\omega}}_i$ are continuous and uniformly bounded because the signals $\{\Delta\dot{\omega}_1, \Delta\dot{\bar{\omega}}_i, \Delta\omega_1, \Delta\bar{\omega}_i, \dot{u}\}$, $i=2,\cdots,N$ are continuous and bounded. Thus, it follows from Theorem B.4 that $\Delta\dot{\omega}_1 \to 0$ and $\Delta\dot{\bar{\omega}}_i \to 0$, $i=2,\cdots,N$, which results in $u_i \to 0$ from (A.23) and (A.24). Using arguments similar to those in the proof for Theorem 5.1, we conclude that q_v converges to the null space of $D\otimes I_3$.

Suppose that $q_v \to 0$. In this case, we note from (5.6) and (5.3) that $\bar{D} \to (-D\otimes I_3)$ and $\bar{D}^T \to (-D^T\otimes I_3)$. On the other hand, since $\dot{q}_v$, $\dot{q}_0$ and $\dot{\tilde{\omega}}$ are all bounded and continuous signals, then from (5.7) we conclude that $\ddot{q}_v$ is also bounded and continuous. From Theorem B.4, we obtain $\dot{q}_v \to 0$. Next we prove $\tilde{\omega} \to 0$. This follows because $\dot{q}_v \to 0$ and because the Jacobian from $\tilde{\omega}$ to $\dot{q}_v$ in (5.7) is always full rank. Therefore, from (5.4) we conclude that ω^B converges to $\mathcal{N}(D^T \otimes I_3)$ since $\bar{D}^T \to -D^T \otimes I_3$. We further note that $\mathcal{N}(D^T \otimes I_3)$ is spanned by $1_N \otimes c$, where $c \in \mathbb{R}^3$ and that ${}^1\omega_1$ converges to $\omega^d(t)$ because $\Delta\omega_1 \to 0$. It follows that $|{}^i\omega_i - \omega^d(t)| \to 0$, which completes the proof.

A.5 Proof of Proposition 8.1

We take the following energy-motivated Lyapunov function

$$V = \sum_{i=1}^{N} P_i(z_i) + \frac{1}{2}(\sum_{i=1}^{N} \xi_i^T m_i \xi_i + \xi_c^T M_c \xi_c) \tag{A.29}$$

where $P_i(z_i)$ is defined in (8.24), and ξ_i and ξ_c are as in (8.13). From Lemma 8.1, the first term in (A.29) is positive definite. Then, the time derivative of V yields

$$\dot{V} = \sum_{i=1}^{N} (f_i - f_i^d)^T \dot{z}_i + \sum_{i=1}^{N} \xi_i^T m_i \ddot{x}_i + \xi_c^T M_c \ddot{x}_c. \tag{A.30}$$

From (8.2), (8.4) and the assumption $\dot{\theta}_c = 0$, the kinematics of z_i are given by

$$\dot{z}_i = \dot{x}_i - \dot{a}_i = \dot{x}_i - \dot{x}_c. \tag{A.31}$$

We next rewrite (A.30) from (8.9), (8.10), (8.11), (8.16) and (A.31) as

$$\begin{aligned}
\dot{V} &= \sum_{i=1}^{N} (f_i - f_i^d)^T (\dot{x}_i - \dot{x}_c) + \sum_{i=1}^{N} \xi_i^T (F_i - f_i) + \xi_c^T \sum_{i=1}^{N} f_i \\
&= \sum_{i=1}^{N} (f_i - f_i^d)^T (\xi_i - \xi_c) + \sum_{i=1}^{N} \xi_i^T (-K_i \xi_i + f_i^d - f_i) + \xi_c^T \sum_{i=1}^{N} f_i \\
&= -\sum_{i=1}^{N} \xi_i^T K_i \xi_i + \xi_c \sum_{i=1}^{N} f_i^d \\
&= -\sum_{i=1}^{N} \xi_i^T K_i \xi_i \le 0,
\end{aligned} \tag{A.32}$$

which implies the stability of the equilibrium $\mathcal{E}$.

To conclude the asymptotic stability, we apply the Invariance Principle and investigate the largest invariant set $\mathcal{M}$ where $\dot{V} = 0$, i.e., $\xi_i = 0$. From (8.13), we note that $\xi_i = 0$ implies that $\dot{x}_i = v$. We further obtain from $\xi_i = 0$ that $\ddot{x}_i = 0$, which leads to $F_i = f_i$ from (8.9). Thus, it is clear from (8.16) that $f_i^d = f_i$. We now show that on $\mathcal{M}$, $\dot{x}_c = v$. To see this, we note that $f_i^d = f_i$ implies that $z_i = z_i^d$. Since z_i^d is constant, we have $\dot{z}_i = 0$ on $\mathcal{M}$, that is, from (8.4), $\dot{x}_i = \dot{a}_i$. Because we consider only the translational motion and because $\dot{x}_i = v$ on $\mathcal{M}$, we conclude that $\dot{a}_i = \dot{x}_c = v$.

A.6 Proof of Proposition 8.2

We choose the following Lyapunov function

$$V_1 = \sum_{i=1}^{N} P_i(z_i) + \frac{1}{2}\sum_{i=1}^{N} \xi_i^T m_i \xi_i + \frac{1}{2}\dot{x}_c M_c \dot{x}_c + \frac{1}{2}\sum_{i=1}^{N} (\hat{v}_i)^T \Lambda_i^{-1} (\hat{v}_i) \tag{A.33}$$

where ξ_i is defined in (8.27). The time derivative of V_1 is given by

$$\dot{V}_1 = -\sum_{i=1}^{N} (f_i^d - f_i)^T \dot{z}_i + \sum_{i=1}^{N} \xi_i^T m_i \dot{\xi}_i + \dot{x}_c^T M_c \ddot{x}_c + \sum_{i=1}^{N} (f_i^d - f_i)^T \dot{\hat{v}}_i. \tag{A.34}$$

Note from (8.28) that

$$\dot{z}_i = \xi_i + \hat{v}_i - \dot{x}_c. \tag{A.35}$$

We then rewrite (A.34) from (8.10), (8.11), (8.29) and (A.35) as

$$\begin{aligned}\dot{V}_1 &= -\sum_{i=1}^{N} (f_i^d - f_i)^T (\xi_i + \hat{v}_i - \dot{x}_c) + \sum_{i=1}^{N} \xi_i^T (-K_i \xi_i + f_i^d - f_i) \\ &\quad + \dot{x}_c^T \sum_{i=1}^{N} f_i + \sum_{i=1}^{N} (f_i^d - f_i)^T \dot{\hat{v}}_i \\ &= -\sum_{i=1}^{N} \xi_i^T K_i \xi_i \leq 0. \end{aligned} \tag{A.36}$$

We next apply the Invariance Principle and investigate the largest invariant set $\mathscr{M}^*$ where $\dot{V}_1 = 0$. On $\mathscr{M}^*$, $\dot{V}_1 = 0$ means $\xi_i = 0$ and thus $\dot{\xi}_i = 0$, which further implies from (8.29) that $f_i = f_i^d$. Then, from (8.25), $\dot{\hat{v}}_i = 0$. It follows from $\ddot{x}_i = \dot{\hat{v}}_i = 0$ that $\dot{x}_i$ is constant on $\mathscr{M}^*$. Since $\dot{f}_i = \dot{f}_i^d = 0$ and $\dot{z}_i = \dot{x}_i - \dot{x}_c$, we conclude $\dot{x}_i = \dot{x}_c$, $\forall i$, which means that all the agents and the payload have the same constant velocity. Noting from (8.27) and $\xi_i = 0$, we further obtain $\dot{x}_c = \hat{v}_i$, $i = 1, \cdots, N$.

Next, from (8.10) and (8.25), we compute

$$M_c \dot{x}_c(t) = \int_0^t \sum_{i=1}^{N} f_i(s) ds + M_c \dot{x}_c(0) \tag{A.37}$$

and

$$\hat{v}_i(t) = \int_0^t \Lambda_i (f_i^d - f_i(s)) ds + \hat{v}_i(0). \tag{A.38}$$

We rewrite (A.38) as

$$\Lambda_i^{-1}(\hat{v}_i(t) - \hat{v}_i(0)) = \int_0^t f_i^d - f_i(s) ds \tag{A.39}$$

and note from (8.11), (A.37) and (A.39) that

$$\sum_{i=1}^{N} \Lambda_i^{-1}(\hat{v}_i(t) - \hat{v}_i(0)) = \int_0^t \sum_{i=1}^{N} (f_i^d - f_i(s))ds \tag{A.40}$$

$$= -M_c(\dot{x}_c(t) - \dot{x}_c(0)). \tag{A.41}$$

Because on $\mathcal{M}^*$, $\dot{x}_c$ and $\hat{v}_i$ are equal and constant, we obtain from (A.41) that $\dot{x}_c = \dot{x}_i = \hat{v}_i = \bar{v}$, where $\bar{v}$ is in (8.32).

A.7 Proof of Corollary 8.3

We consider the Lyapunov function below

$$V_a = \sum_{i=1}^{N} P_i(z_i) + \frac{1}{2}\sum_{i=1}^{N} \xi_i^T m_i \xi_i + \frac{1}{2}\xi_c^T M_c \xi_c + \frac{1}{2}\sum_{i=2}^{N} (\tilde{v}_i)^T \Lambda_i^{-1} \tilde{v}_i, \tag{A.42}$$

where ξ_c is in (8.13), $\xi_1 = \dot{x}_1 - v$, ξ_i, $i = 2,\cdots,N$, are defined in (8.27) and $\tilde{v}_i$, $i = 2,\cdots,N$ are as in (8.33). The time derivative of V_a is computed from (8.10), (8.13), (8.11), (8.25), (8.28) and (8.29) as

$$\begin{aligned}\dot{V}_a = &-\sum_{i=1}^{N} (f_i^d - f_i)^T (\xi_i + \hat{v}_i - v - \xi_c) + \sum_{i=1}^{N} \xi_i^T (-K_i \xi_i + f_i^d - f_i) \\ &+ \xi_c^T \sum_{i=1}^{N} f_i + \sum_{i=1}^{N} (f_i^d - f_i)^T \tilde{v}_i \\ = &-\sum_{i=1}^{N} \xi_i^T K_i \xi_i \leq 0\end{aligned} \tag{A.43}$$

which implies the stability of $\mathcal{E}_a^*$ in (8.34). We perform a Lyapunov analysis similar to the proof in Appendix A.6 and conclude the global asymptotic stability of $\mathcal{E}_a^*$ in (8.34).

A.8 Proof of Theorem 9.1

We define the new variable

$$\zeta := (S \otimes I_p)\mathbf{x}, \tag{A.44}$$

where S is as in (9.9) and, thus, $\mathbf{x}_i - \mathbf{x}_j \to 0$, $\forall i, j$, is equivalent to $\zeta \to 0$. To prove asymptotic stability of $\zeta = 0$, we note from (9.5)-(9.6) that

$$\begin{aligned}\dot{\zeta} &= -(SD(t) \otimes I_p)\psi((D(t)^T \otimes I_p)\mathbf{x}) \\ &= -(QS(t) \otimes I_p)\psi((D(t)^T S^T \otimes I_p)\zeta)\end{aligned} \tag{A.45}$$

where we obtained the second equation by substituting

$$D(t)^T = D(t)^T S^T S \tag{A.46}$$

and by using (A.44). To see that (A.46) holds, note that $I_N - S^T S$ is an orthogonal projection matrix onto the span of 1_N, and that 1_N is in the null space of $D(t)^T$ for all $t \geq 0$ which, together, imply

$$D(t)^T (I_N - S^T S) = 0 \quad \Rightarrow \quad D(t)^T = D(t)^T S^T S. \tag{A.47}$$

We then denote

$$F(t) := SD(t) \otimes I_p \tag{A.48}$$

and conclude global uniform asymptotic stability for (A.45) from Lemma A.1 below.

Lemma A.1. Consider the time-varying system

$$\dot{\zeta} = -F(t)\psi(F(t)^T \zeta) \tag{A.49}$$

where $\zeta \in \mathbb{R}^n$, $F(t)$ is an $n \times r$ piecewise continuous matrix of t, and $\psi : \mathbb{R}^r \to \mathbb{R}^r$ is a locally Lipschitz nonlinearity satisfying (2.46). If $F(t)$ satisfies $\|F(t)\| \leq \mu$ and

$$\int_t^{t+\delta} F(\tau)F(\tau)^T d\tau \geq \alpha I \tag{A.50}$$

for some constants μ, δ and α that do not depend on t, then the origin $\zeta = 0$ is globally uniformly asymptotically stable. □

Proof of Lemma A.1: We let

$$\phi(t, \zeta) := F(t)^T \zeta \tag{A.51}$$

and note that the Lyapunov function

$$V_1(\zeta) = \frac{1}{2}\zeta^T \zeta \tag{A.52}$$

satisfies

$$\dot{V}_1 = -\phi^T \psi(\phi) =: Y_1(\phi) \leq 0, \tag{A.53}$$

from which we conclude global uniform stability. To prove global uniform asymptotic stability we employ the Nested Matrosov Theorem in Appendix B.5. To this end we introduce the auxiliary function

$$V_2(t, \zeta) = -\zeta^T S(t)\zeta \qquad S(t) := \int_t^{\infty} e^{(t-\tau)} F(\tau)F(\tau)^T d\tau \tag{A.54}$$

where

$$\|S(t)\| \leq \mu^2 \tag{A.55}$$

from $\|F(t)\| \leq \mu$, and

$$S(t) \geq \int_t^{t+\delta} e^{(t-\tau)} F(\tau)F(\tau)^T d\tau \geq \alpha e^{-\delta} I \tag{A.56}$$

from (A.50). Furthermore, $\dot{S}(t) = S(t) - F(t)F(t)^T$, from which we obtain

$$\dot{V}_2 \leq -\zeta^T S(t)\zeta + \phi^T\phi - 2\zeta^T S(t)\dot{\zeta}. \tag{A.57}$$

Next, substituting (A.55)-(A.56) and $|\dot{\zeta}| \leq \mu|\psi(\phi)|$ obtained from (A.49), we get

$$\dot{V}_2 \leq -\alpha e^{-\delta}|\zeta|^2 + |\phi|^2 + 2\mu^3|\zeta||\psi(\phi)| =: Y_2(\zeta,\phi). \tag{A.58}$$

When $Y_1(\phi) = 0$ in (A.53), it follows from (2.46) that $\phi = 0$ and, thus, $Y_2(\zeta,\phi) = -\alpha e^{-\delta}|\zeta|^2 \leq 0$. Furthermore, $Y_1(\phi) = 0$ and $Y_2(\zeta,\phi) = 0$ together imply $\zeta = 0$, which means that all conditions of the Nested Matrosov Theorem are satisfied and, hence, $\zeta = 0$ is globally uniformly asymptotically stable.

Appendix B
Technical Tools Used in the Book

B.1 Schur Decomposition

The Schur decomposition transforms a square matrix into an upper triangular matrix using a unitary similarity transformation.

Theorem B.1. *[15, Theorem 5.4.1]*

Let $L \in \mathbb{C}^{n\times n}$. Then, there exist a unitary matrix $Q \in \mathbb{C}^{n\times n}$ and an upper triangular matrix $B \in \mathbb{C}^{n\times n}$ such that

$$L = QBQ^{-1}. \tag{B.1}$$

□

Because B is similar to L, B has the same eigenvalues with the same algebraic multiplicities as L. Because B is upper triangular, these eigenvalues are the diagonal entries of B.

B.2 Invariance Principle [69, Theorem 4.4]

Theorem B.2. *Let $\Omega \subset D$ be a compact set that is positively invariant with respect to the following dynamical system*

$$\dot{x} = f(x) \tag{B.2}$$

where $f : D \to \mathbb{R}^n$ is a locally Lipschitz map from a domain $D \in \mathbb{R}^n$ into $\mathbb{R}^n$. Let $V : D \to \mathbb{R}$ be a continuously differentiable function such that $\dot{V}(x) \le 0$ in Ω. Let E be the set of all points in Ω where $\dot{V}(x) = 0$. Let M be the largest invariant set in E. Then every solution starting in Ω approaches M as $t \to \infty$. □

B.3 Barbalat's Lemma

Theorem B.3 (Barbalat's Lemma).
Let $f(t) : \mathbb{R} \to \mathbb{R}$ be a uniformly continuous function on $[0,\infty)$. Suppose that $\lim_{t\to\infty} \int_0^t f(s)ds$ exists and is finite. Then

$$f(t) \to 0 \quad \text{as} \quad t \to \infty. \tag{B.3}$$

□

Barbalat's Lemma leads to the following convergence result:

Theorem B.4. *Let $\xi(t) : \mathbb{R} \to \mathbb{R}$ be a continuous function defined on $[0,\infty)$. If $\xi(t) \to 0$ as $t \to \infty$ and $\ddot{\xi}(t)$ is bounded, then $\dot{\xi}(t) \to 0$ as $t \to \infty$.* □

Proof. Let $\dot{\xi}(t)$ be the $f(t)$ in Theorem B.3. Then $\lim_{t\to\infty} \int_0^t f(s)ds = \lim_{t\to\infty} \xi(t) - \xi(0) = -\xi(0)$. Because $\ddot{\xi}$ is bounded, $\dot{f}(t)$ is bounded and thus $f(t)$ is uniformly continuous. The assumptions in Theorem B.3 are satisfied and it follows that $f(t) = \dot{\xi}(t) \to 0$ as $t \to \infty$. □

When applying Barbalat's Lemma to the Lyapunov analysis of nonautonomous systems, we have the following result:

Theorem B.5. *Suppose $f(t,x)$ is piecewise continuous in t and locally Lipschitz in x, uniformly in t, on $[0,\infty) \times \mathbb{R}^n$. Furthermore, suppose $f(t,0)$ is uniformly bounded for all $t \geq 0$. Let $V : [0,\infty) \times \mathbb{R}^n \to R$ be a continuously differentiable function such that*

$$W_1(x) \leq V(t,x) \leq W_2(x) \tag{B.4}$$

$$\dot{V}_1(x) = \frac{\partial V}{\partial t} + \frac{\partial V}{\partial x} f(t,x) \leq -W(x) \tag{B.5}$$

$\forall t \geq 0$, $\forall x \in \mathbb{R}^n$, where $W_1(x)$ and $W_2(x)$ are continuous positive definite functions and $W(x)$ is a continuous positive semidefinite function on $\mathbb{R}^n$. Then, all solutions of $\dot{x} = f(t,x)$ are bounded and satisfy

$$W(x(t)) \to 0 \quad as \quad t \to 0. \tag{B.6}$$

□

B.4 Proposition 2.44 in [119]

Proposition B.1 (Relative degree of nonlinear passive systems).
Consider an input-affine nonlinear system

$$\begin{aligned} \dot{x} &= f(x) + g(x)u \\ y &= h(x). \end{aligned} \tag{B.7}$$

If (B.7) is passive with a C^2 storage function $S(x)$, then it has relative degree one at $x = 0$. □

The result in this proposition means that the matrix $L_g h(0) := \frac{\partial h(x)}{\partial x}\big|_{x=0} g(0)$ is invertible if (B.7) is passive.

B.5 Nested Matrosov Theorem [85, Theorem 1]

Theorem B.6 (Nested Matrosov Theorem).
Consider a dynamical system

$$\dot{x} = F(t,x), \quad x \in \mathbb{R}^n. \tag{B.8}$$

For $\Delta \geq 0$, define $\mathscr{B}(\Delta) := \{x \in \mathbb{R}^n | 0 \leq |x| \leq \Delta\}$. Under the following assumptions, the origin of (B.8) is uniformly globally asymptotically stable.

Assumption 1: *The origin of (B.8) is uniformly globally stable.*

Assumption 2: *There exist integers $j, m > 0$ and for each $\Delta > 0$ there exist*

- *a number $\mu > 0$;*
- *locally Lipschitz continuous functions $V_i : \mathbb{R}_{\geq 0} \times \mathbb{R}^n \to \mathbb{R}$, $i \in \{1, \cdots, j\}$;*
- *a function $\phi : \mathbb{R}_{\geq 0} \times \mathbb{R}^n \to \mathbb{R}^m$;*
- *continuous functions $Y_i : \mathbb{R}^n \times \mathbb{R}^m \to \mathbb{R}$, $i \in \{1, \cdots, j\}$;*

such that, for almost all $(t,x) \in \mathbb{R}_{\geq 0} \times \mathscr{B}(\Delta)$, and all $i \in \{1, \cdots, j\}$

$$\max\{|V_i(t,x), \phi(t,x)|\} \leq \mu \tag{B.9}$$

$$\dot{V}_i(t,x) \leq Y_i(x, \phi(t,x)). \tag{B.10}$$

Assumption 3: *For each integer $k \in \{1, \cdots, j\}$, we have that*

A) $\{Y_i(z, \psi) = 0, \forall i \in \{1, \cdots, k-1\}$, and all $(z, \psi) \in \mathscr{B}(\Delta) \times \mathscr{B}(\mu)\}$ implies that

B) $\{Y_k(z, \psi) \leq 0$, for all $(z, \psi) \in \mathscr{B}(\Delta) \times \mathscr{B}(\mu)\}$.

Assumption 4: *We have that the statement*

A) $\{Y_i(z, \psi) = 0, \forall i \in \{1, \cdots, j\}$, and all $(z, \psi) \in \mathscr{B}(\Delta) \times \mathscr{B}(\mu)\}$ implies that

B) $\{z = 0\}$. □

B.6 Lemma 4.7 in [69]

Consider the cascade system

$$\dot{x}_1 = f_1(t, x_1, x_2) \tag{B.11}$$

$$\dot{x}_2 = f_2(t, x_2) \tag{B.12}$$

where $f_1 : [0,\infty) \times \mathbb{R}^{n_1} \times \mathbb{R}^{n_2} \to \mathbb{R}^{n_1}$ and $f_2 : [0,\infty) \times \mathbb{R}^{n_2} \to \mathbb{R}^{n_2}$ are piecewise continuous in t and locally Lipschitz in $x = \begin{pmatrix} x_1 \\ x_2 \end{pmatrix}$.

Theorem B.7. *Under the stated assumptions, if the system (B.11), with x_1 as input, is Input-to-State Stable (ISS) and the origin of (B.12) is globally uniformly asymptotically stable, then the origin of the cascade system (B.11) and (B.12) is globally uniformly asymptotically stable.* □

B.7 Theorem 4.19 in [69]

Theorem B.8. *Consider the system*

$$\dot{x} = f(t,x,u) \tag{B.13}$$

where $f : [0,\infty) \times \mathbb{R}^n \times \mathbb{R}^m \to \mathbb{R}^n$ is piecewise continuous in t and locally Lipschitz in x and u. Let $V : [0,\infty) \times \mathbb{R}^n \to \mathbb{R}$ be a continuous differentiable function such that

$$\alpha_1(|x|) \leq V(t,x) \leq \alpha_2(|x|) \tag{B.14}$$

$$\frac{\partial V}{\partial t} + \frac{\partial V}{\partial x} f(t,x,u) \leq -W_3(x), \quad \forall |x| \geq \rho(|u|) > 0 \tag{B.15}$$

$\forall (t,x,u) \in [0,\infty) \times \mathbb{R}^n \times \mathbb{R}^m$, where α_1, α_2 are class K_∞ functions, ρ is a class K function and $W_3(x)$ is a continuous positive definite function on $\mathbb{R}^n$. Then the system (B.13) is input-to-state stable. □

B.8 Proposition 2 in [65]

Proposition B.2. *Assume a storage function $H(x)$ is strictly convex. Then, for every $x^* \in \mathbb{R}^n$, the new storage function $H_0(x)$ defined as*

$$H_0(x) = H(x) - H(x^*) - (x - x^*)^T \nabla H(x^*) \tag{B.16}$$

has a unique global minimum at x^* and is proper. □

B.9 Theorem 10.4 in [69]

This theorem is a combination of Theorem 10.4 in [69] and the discussion right after that theorem.

Theorem B.9. *Let $f(t,x,\varepsilon)$ and its partial derivatives with respect to (x,ε) up to the second order be continuous and bounded for $(t,x,\varepsilon) \in [0,\infty) \times D_0 \times [0,\varepsilon_0]$, for every compact set $D_0 \subset D$, where $D \subset \mathbb{R}^n$ is a domain. Suppose f is T-periodic in t for some $T > 0$ and ε is a positive parameter. Let $x(t,\varepsilon)$ and $x_{\text{av}}(\varepsilon t)$ denote the solutions of*

$$\dot{x} = \varepsilon f(t,x,\varepsilon) \tag{B.17}$$

and

$$\dot{x} = \varepsilon f_{\text{av}}(x), \tag{B.18}$$

respectively, where

$$f_{\text{av}}(x) = \frac{1}{T}\int_0^T f(\tau,x,0)d\tau. \tag{B.19}$$

If the origin $x = 0 \in D$ is an exponentially stable equilibrium point of the average system (B.18) and $f(t,0,\varepsilon) = 0$ for all $(t,\varepsilon) \in [0,\infty) \times [0,\varepsilon_0]$, then there exists positive constants ε^* and k such that, for all $0 < \varepsilon < \varepsilon^*$, the origin is an exponentially stable equilibrium of (B.17). □

B.10 Theorem 3.4.11 in [62]

We only cite the relevant part of [62, Theorem 3.4.11] below.

Theorem B.10. *Consider a linear time-varying dynamical system*

$$\dot{x}(t) = A(t)x(t), \quad x(t) \in \mathbb{R}^n. \tag{B.20}$$

Let the elements of $A(t)$ in (B.20) be differentiable and bounded function of time and assume that $Re\{\lambda_i(A(t))\} \leq -\sigma_s \ \forall t \ \forall i = 1,\cdots,n$, where $\lambda_i(A(t))$ is the ith eigenvalue of $A(t)$ and $\sigma_s > 0$ is some constant.

If $||\dot{A}|| \leq \mu$ is satisfied for some positive μ and $\forall t \geq 0$, then there exists a $\mu^* > 0$ such that if $\mu \in [0,\mu^*)$, the equilibrium state x_e of (B.20) is globally uniformly asymptotically stable. □

B.11 Summary of Example 11.14 in [69]

In [69, Example 11.14], the following system

$$\dot{x} = f(t,x,v) \tag{B.21}$$

$$\varepsilon\dot{z} = Az + Bu \tag{B.22}$$

$$v = Cz \tag{B.23}$$

is considered, in which $f(t,0,0) = 0$ and A is a Hurwitz matrix. Assume that $-CA^{-1}B = I$. Setting ε in (B.22) to zero leads to the reduced model

$$\dot{x} = f(t,x,u). \tag{B.24}$$

Let $u = \gamma(t,x)$ be a state feedback control law such that the origin of (B.24) is exponentially stable. Assume that f and γ is sufficiently smooth. Then the origin of the original system (B.21)-(B.23) is exponentially stable for sufficiently small ε.

B.12 Rigid Body Attitude and Its Parameterizations

This section presents the basic language and tools used to describe the rotational motion of a rigid body. Recall that a body is rigid if the distance between any two points fixed with respect to the body remains constant. If the body is free to move and rotate in space, it has 6 degree-of-freedom (DOF), 3 translational and 3 rotational. When restricted to plane, the body has 3 DOF, 2 translational and 1 rotational. This section will mainly focus on the rotational motion of a rigid body.

B.12.1 Rigid Body Attitude

As shown in Fig. B.1, the *attitude* of a rigid body in $\mathbb{E}^3$ (Euclidean 3-space) is completely characterized by an orthonormal frame R attached to the body. Define a right-handed orthonormal frame as follows:

Definition B.1. $\mathrm{R} = (\mathbf{r_1}\ \ \mathbf{r_2}\ \ \mathbf{r_2})$ is an orthonormal frame if

1. $||\mathbf{r_i}|| = 1$, $i = 1,2,3$ (normality)
2. $\mathbf{r_i} \cdot \mathbf{r_j} = 0$ if $i \neq j$, $i,j = 1,2,3$ (orthogonality)
3. $\mathbf{r_1} \times \mathbf{r_2} = \mathbf{r_3}$ (right handed rule)

where $\mathbf{r_i}$'s are in $\mathbb{E}^3$. □

Given an inertial frame E, we can write the coordinates of $\mathbf{r_i}$'s in E as $r_i \in \mathbb{R}^3$, that is, $r_i = [r_{i,1}, r_{i,2}, r_{i,3}]^T$. Then the attitude matrix R of the body in E is given by

$$R = [r_1\ \ r_2\ \ r_3]. \tag{B.25}$$

The attitude matrix R satisfies $R^T R = I_3$ (due to orthonormality of r_i's) and $\det R = 1$ (due to the right handed rule) and thus belongs to the Lie Group $SO(3)$ (Special Orthogonal group of dimension 3) defined as

$$SO(3) = \left\{R \in \mathbb{R}^{3\times 3} | R^T R = I_3,\ \ \det R = 1\right\}. \tag{B.26}$$

The matrix R is used not only for representing the attitude of a rigid body in the inertial frame E, but also for rotating a vector from the inertial frame to the body frame or vice versa. Given a vector $\mathbf{v} \in \mathbb{E}^3$, let its coordinates in the inertial frame E be $v_E \in \mathbb{R}^3$. Then its coordinates in the body frame v_B are given by

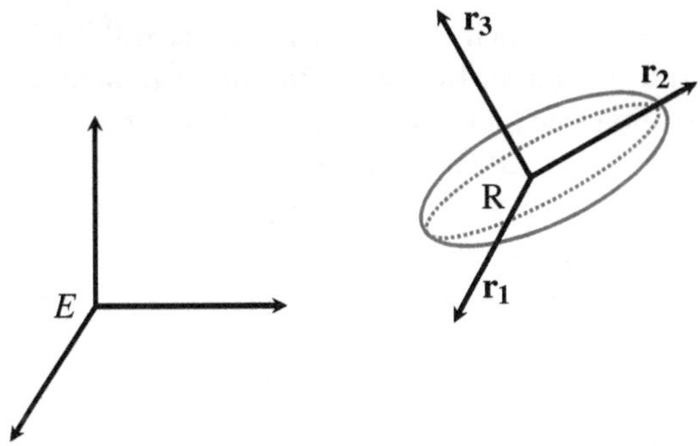

Fig. B.1 The attitude of a rigid body is characterized by a frame R $= (\mathbf{r}_1, \mathbf{r}_3, \mathbf{r}_3)$. If we represent R in an inertial frame E, we obtain an attitude matrix R as in (B.25).

$$v_B = R^T v_E. \tag{B.27}$$

B.12.2 Parameterizations of Attitude Matrix

Note that there are 9 parameters in the attitude matrix R, subject to 6 constraints imposed by the orthogonality. For manipulation, analysis and implementation, it is more convenient to use other representations of $SO(3)$. The minimal number of parameters that represents $SO(3)$ is 3 since there are 3 free parameters in the attitude matrix. We will present a class of three-parameter representations of $SO(3)$, including Euler-Rodriguez parameters, Modified Gibb's vector, and vector quaternion. This class of parameterizations all consists of the equivalent axis and angle defined by the Euler Rotation Theorem below:

Theorem B.11 (Euler Rotation Theorem). *Every rigid body rotation is equivalent to rotating around an equivalent axis over an angle.* □

Let the equivalent axis of R be represented by a unit vector $h \in \mathbb{R}^3$ and the corresponding rotation angle be θ. We consider three-parameter parameterizations of the following form:

$$s = \gamma(\theta)h \tag{B.28}$$

where $\gamma(\cdot) : \mathbb{R} \to \mathbb{R}$ is a C^1 odd function. Included in this general parameterization (B.28) are

$$\gamma(\theta) = \tan(\frac{\theta}{2}) \quad \text{(Euler-Rodriguez parameter, Gibb's Vector)} \tag{B.29}$$

$$\gamma(\theta) = \tan(\frac{\theta}{4}) \quad \text{(Modified Gibb's Vector)} \tag{B.30}$$

$$\gamma(\theta) = \sin(\frac{\theta}{2}) \quad \text{(Vector quaternion).} \tag{B.31}$$

Other parameterizations, such as Euler angles and roll-pitch-yaw angles, are also commonly used in literature. However, only the parameterization in (B.28) is discussed in this section since it is closely related to Chapter 5.

The representation (B.28) can be uniquely determined for almost all $R \in SO(3)$. Since (B.28) relates to the equivalent axis/angle, we first present how to obtain (h, θ) from a given $R \in SO(3)$.

The relationship between the rotation matrix R and (h, θ) is given by the Rodriguez formula

$$R = I_3 + \sin\theta \hat{h} + (1 - \cos\theta)(\hat{h})^2 \tag{B.32}$$

where $\hat{}$ is defined in (1.5). We obtain from (B.32) that

$$\cos\theta = \frac{1}{2}(\mathrm{tr}(R) - 1) \tag{B.33}$$

and

$$\sin\theta = \frac{1}{2} h^T (R - R^T)^\vee \tag{B.34}$$

which implies

$$\theta = \mathrm{atan2}(\frac{1}{2} h^T (R - R^T)^\vee, \frac{1}{2}(\mathrm{tr}(R) - 1)) \tag{B.35}$$

where $\mathrm{tr}(R)$ is the trace of R and $^\vee$ is the inverse operation of $\hat{}$ defined in (1.6). By manipulating the diagonal elements of R, we can find an expression for the ith element of h:

$$h_i = \pm \left(\frac{2R_{ii} - \mathrm{tr}(R) + 1}{3 - \mathrm{tr}(R)} \right)^{\frac{1}{2}} \tag{B.36}$$

where R_{ij} denotes the (i, j)th element of R. The computation of (B.36) breaks down when $\mathrm{tr}(R) = 3$, that is, $R = I_3$, which implies from (B.33) that $\theta = 0$. In this case, h can be an arbitrary unit vector. Once h is obtained, we compute θ from (B.35).

Note that (B.36) gives two solutions to h, which results in two solutions to (B.35). These two solutions correspond to the two possible direction of rotation to achieve the same end result. We can also verify from (B.32) that both (θ, h) and $(-\theta, -h)$ yield the same R.

The parameterization (B.28) can be readily obtained from h and θ. Thanks to the odd function $\gamma(\cdot)$, (θ, h) and $(-\theta, -h)$ yield the same parameterization s. Thus, for almost all $R \in SO(3)$, one can uniquely determine a three parameterization s from (θ, h) and (B.28). In certain cases, for example when $\theta = \pi$ and $\gamma(\theta) = \tan\frac{\theta}{2}$, (B.28) is not well defined. This is because all three-parameter representations of $SO(3)$ are only locally one-to-one and onto. As we will present in the next section, there are always singularities for three-parameter representations of $SO(3)$. This singularity issue can be eliminated by using unit quaternion, which uses the least possible number of parameters (4) to represent the orientation globally. Besides, unit quaternion is of great interest due to its desirable computational properties [144].

A unit quaternion q consists of a scalar part q_0 and a vector part $q_v \in \mathbb{R}^3$, which can be defined through (θ, h) as

$$q_0 = \cos\frac{\theta}{2} \quad \text{and} \quad q_v = h\sin\frac{\theta}{2}. \tag{B.37}$$

It follows that $|q| = q_0^2 + q_v^T q_v = 1$. Thus, q is a four-parameter representation of $SO(3)$ with one constraint. Given q, the corresponding attitude matrix R can be obtained from

$$R = I_3 + 2q_0\widehat{q_v} + 2\widehat{q_v}^2. \tag{B.38}$$

It is easy to verify from (B.38) that both q and $-q$ correspond to the same attitude matrix R.

Conversely, given $R \in SO(3)$, we obtain from (B.33) that

$$q_0 = \cos\frac{\theta}{2} = \pm\sqrt{\frac{1+\cos\theta}{2}} = \pm\frac{1}{2}\sqrt{1+\mathrm{tr}(R)}. \tag{B.39}$$

When $q_0 \neq 0$, we find

$$q_v = \left(\frac{R - R^T}{4q_0}\right)^{\vee} \tag{B.40}$$

because (B.38) implies $R - R^T = 4q_0\widehat{q_v}$. When $q_0 = 0$, q_v can be found as the unit eigenvector of R corresponding to the eigenvalue 1 since $Rq_v = q_v$.

B.13 Rigid Body Kinematics

The rotation kinematics of rigid body describes the evolution of the attitude of a rigid body. Let the attitude matrix of a rigid body with respect to an inertial frame at time t be $R(t)$. Using the identity $R(t)R^T(t) = I_3$, we obtain

$$\dot{R}(t)R^T(t) + R(t)\dot{R}^T(t) = 0 \tag{B.41}$$

which implies $\dot{R}(t)R^T(t)$ is skew symmetric. Defining

$$\widehat{\omega} = \dot{R}(t)R^T(t) \tag{B.42}$$

we get

$$\dot{R} = \widehat{\omega}R \tag{B.43}$$

where $\omega \in \mathbb{R}^3$ is defined as the angular velocity of $R(t)$ represented in the inertial frame. If we denote by ω^B the angular velocity of $R(t)$ represented in the body frame $R(t)$, we obtain from (B.27)

$$\omega^B = R^T\omega \tag{B.44}$$

and

$$\dot{R} = \widehat{R\omega^B}R = R\widehat{\omega^B} \tag{B.45}$$

where we used the fact

$$\widehat{R\omega^B} = R\widehat{\omega^B}R^T. \tag{B.46}$$

Let $s \in \mathbb{R}^m$ be a parameterization of $R \in SO(3)$. Then the time derivative of s is given by

$$\dot{s} = J_s(s)\omega \tag{B.47}$$

where $J_s(s)$ is called the representation Jacobian matrix. If $J_s(s)$ loses rank, s cannot be computed from (B.51). This is called a singularity of the representation s.

We now compute the Jacobian matrices of the parameterizations in (B.28). First, we use (B.33) and (B.43) to obtain

$$\dot{\theta} = h^T \omega. \tag{B.48}$$

We further calculate $\dot{h}$ from (B.32) as

$$\dot{h} = -\frac{1}{2}\left(\widehat{h} + \cot\frac{\theta}{2}\widehat{h}^2\right)\omega. \tag{B.49}$$

Note from (B.28) that

$$\dot{s} = \frac{\partial \gamma}{\partial \theta}\dot{\theta} h + \gamma(\theta)\dot{h}. \tag{B.50}$$

It then follows from (B.48) and (B.49) that

$$\dot{s} = \left[\frac{\partial \gamma}{\partial \theta} hh^T - \frac{1}{2}\gamma(\theta)\left(\widehat{h} + \cot\frac{\theta}{2}\widehat{h}^2\right)\right]\omega. \tag{B.51}$$

Using the identities $\widehat{h}^2 = hh^T - h^T h I_3$ and $|h| = 1$, we rewrite (B.51) as

$$\dot{s} = \underbrace{\left[\left(\frac{\partial \gamma}{\partial \theta} - \frac{1}{2}\gamma(\theta)\cot\frac{\theta}{2}\right)hh^T - \frac{1}{2}\gamma(\theta)\widehat{h} - \frac{1}{2}\gamma(\theta)\cot\frac{\theta}{2}I_3\right]}_{J_s(s)}\omega. \tag{B.52}$$

Thus, s is singular whenever $J_s(s)$ is singular or unbounded. For example, when $\gamma(\theta) = \sin\frac{\theta}{2}$ (i.e., s is the vector quaternion), $J_s(s)$ loses rank if $\cos\frac{\theta}{2} = 0$, which implies that $s^T s = \gamma(\theta)^2 h^T h = 1$. If s is the Gibb's vector, $\gamma(\theta) = \tan\frac{\theta}{2}$ and $J_s(s)$ becomes unbounded at $\theta = \pi$, which means that s is unbounded. For unit quaternion in (B.37), we compute its time derivative as

$$\begin{pmatrix} \dot{q}_0 \\ \dot{q}_v \end{pmatrix} = \underbrace{\frac{1}{2}\begin{pmatrix} -q_v^T \\ q_0 I_3 - \widehat{q}_v \end{pmatrix}}_{J_q(q)}\omega. \tag{B.53}$$

Note that $J_q(q)^T J_q(q) = \frac{1}{4}(1 + q_0^2)I_3$, which is never singular. For this reason, unit quaternion is a popular representation of $SO(3)$.

B.14 Rigid Body Dynamics

When a rigid body rotates at the angular velocity ω in the inertial frame, the linear velocity of a point p on the rigid body is given by $\widehat{\omega}p$. Let $\rho(p)$ be the density of the body at point p. Then the total rotational energy of this rigid body is given

$$K = \frac{1}{2}\int_V \rho(p)(\widehat{\omega}p)^T(\widehat{\omega}p)dV \tag{B.54}$$

which is equivalent to

$$K = \frac{1}{2}\omega^T \int_V \rho(p)(p^T p I_3 - pp^T)dV\,\omega. \tag{B.55}$$

Defining

$$\mathscr{I} = \int_V \rho(p)(p^T p I_3 - pp^T)dV, \tag{B.56}$$

we have

$$K = \frac{1}{2}\omega^T \mathscr{I}\omega. \tag{B.57}$$

The matrix $\mathscr{I}$ is the inertia matrix of the rigid body represented in the inertial frame. It is always positive definite. If the body is rotating, $\mathscr{I}$ also changes.

The angular momentum of the body is given by $M = \mathscr{I}\omega$ in the inertial frame. The time derivative of M is the torque τ acting on the body represented in the inertial frame, i.e.,

$$\frac{dM}{dt} = \frac{d(\mathscr{I}\omega)}{dt} = \tau. \tag{B.58}$$

As the body rotates, the density $\rho(p)$ does not change. Therefore, using $\dot{p} = \widehat{\omega}p$, we obtain from (B.56) that

$$\frac{d\mathscr{I}}{dt} = \widehat{\omega}\mathscr{I} - \mathscr{I}\widehat{\omega}. \tag{B.59}$$

It then follows that

$$\mathscr{I}\dot{\omega} + \widehat{\omega}\mathscr{I}\omega = \tau, \tag{B.60}$$

which is the Euler equation.

We can also write (B.60) in the body frame. We use (B.44) and (B.45) in (B.60) and obtain

$$\mathscr{I}\frac{d(R\omega^B)}{dt} + \widehat{R\omega^B}\mathscr{I}R\omega^B = \tau \tag{B.61}$$

which can be simplified as

$$\mathscr{I}^B\dot{\omega}^B + \widehat{\omega^B}\mathscr{I}^B\omega^B = \tau^B, \tag{B.62}$$

where $\tau^B = R^T\tau$ is the torque τ represented in the body frame and

$$\mathscr{I}^B = R^T\mathscr{I}R \tag{B.63}$$

is the inertia matrix represented in the body frame. Note that (B.60) and (B.62) are of the same form.

Index

References

1. Agrachev, A.A., Liberzon, D.: Lie-algebraic stability criteria for switched systems. SIAM Journal on Control and Optimization **40**(1), 253–269 (2001)
2. Aguiar, A.P., Hespanha, J.P., Kokotović, P.V.: Path-following for non-minimum phase systems removes performance limitations. IEEE Transactions on Automatic Control **50**(2), 234–239 (2005)
3. Al-Hiddabi, S.A., McClamroch, N.H.: Tracking and maneuver regulation control for nonlinear nonminimum phase systems: Application to flight control. IEEE Transactions on Control Systems Technology **10**(6), 780–792 (2002)
4. Anderson, B.D.O., Dasgupta, S., Yu, C.: Control of directed formations with a leader-first follower structure. In: Proceedings of the 46th IEEE Conference on Decision and Control (2007)
5. Arcak, M.: Passivity As a Design Tool for Group Coordination. IEEE Transactions on Automatic Control **52**(8), 1380–1390 (2007)
6. Arimoto, S., Miyazaki, F.: On the stability of PID feedback with sensory information. In: M. Brady, R. Paul (eds.) Robotics Research. MIT Press (1984)
7. Ariyur, K.B., Krstić, M.: Real-Time Optimization by Extremum-Seeking Feedback. Wiley-Interscience, Hoboken, NJ (2003)
8. Bachmayer, R., Leonard, N.: Vehicle networks for gradient descent in a sampled environment. In: the 41st IEEE Conference on Decision and Control, pp. 113–117. Las Vegas, NV (2002)
9. Bai, H.: The undesired equilibria of formation control with ring graphs. In: Proceedings of the American Control Conference. Baltimore, MD (2010)
10. Bai, H., Arcak, M., Wen, J.T.: Rigid body attitude coordination without inertial frame information. Automatica, **44**(12), 3170–3175 (2008)
11. Bai, H., Arcak, M., Wen, J.T.: Adaptive design for reference velocity recovery in motion coordination. Systems & Control Letters **57**(8), 602–610 (2008)
12. Bai, H., Arcak, M., Wen, J.T.: Using orientation agreement to achieve planar rigid formation. In: Proceedings of the American Control Conference, pp. 753–758. Seattle, WA (2008)
13. Bai, H., Arcak, M., Wen, J.T.: Adaptive motion coordination: Using relative velocity feedback to track a reference velocity. Automatica **45**(4), 1020–1025 (2009)
14. Bai, H., Wen, J.T.: Cooperative Load Transport: A Formation-Control Perspective. IEEE Transactions on Robotics **26**(4), 742–750 (2010)
15. Bernstein, D.S.: Matrix mathematics: theory, facts, and formulas, second edn. Princeton University Press (2009)
16. Bertsekas, D., Tsitsiklis, J.N.: Parallel and Distributed Computation: Numerical Methods. Prentice Hall (1989)
17. Biggs, N.: Algebraic Graph Theory, second edn. Cambridge University Press (1993)
18. Biyik, E., Arcak, M.: Gradient climbing in formation via extremum seeking and passivity-based coordination rules. Asian Journal of Control (Special Issue on Collective Behavior and Control of Multi-Agent Systems) **10**(2), 201–211 (2008)
19. B.Jayawardhana, G.Weiss: Disturbance rejection with LTI internal models for passive nonlinear systems. In: Proceedings of the 16th IFAC World Congress (2005)
20. Bondhus, A.K., Pettersen, K.Y., Gravdahl, J.T.: Leader/follower synchronization of satellite attitude without angular velocity measurements. In: Proceedings of the 44th IEEE Conference on Decision and Control, pp. 7270–7277. Seville, Spain (2005)
21. Byrnes, C.I., Priscoli, F.D., Isidori, A.: Output Regulation of Uncertain Nonlinear Systems. Birkhäuser, Boston (1997)
22. Cao, M., Morse, A.S., Yu, C., Anderson, B.D.O., Dasgupta, S.: Controlling a triangular formation of mobile autonomous agents. In: Proceedings of the 46th IEEE Conference on Decision and Control, pp. 3603–3608. New Orleans, LA, USA (2007)
23. Cao, M., Yu, C., Morse, A.S., Anderson, B.D.O., Dasgupta, S.: Generalized controller for directed triangle formations. In: Proceedings of the 17th World Congress of The International Federation of Automatic Control, pp. 6590–6595. Seoul, Korea (2008)

24. Cao, Y., Li, Y., Ren, W., Chen, Y.: Distributed coordination algorithms for multiple fractional-order systems. In: Proceedings of the IEEE Conference on Decision and Control, pp. 2920–2925. Cancun, Mexico (December 2008)
25. Casbeer, D., Kingston, D., Beard, R., Li, S.M., McLain, T., Mehra, R.: Cooperative forest fire surveillance using a team of small unmanned air vehicles. International Journal of Systems Sciences **37**(6), 351–360 (2006)
26. Chen, Y.Q., Wang, Z.: Formation control: a review and a new consideration. In: IEEE/RSJ International Conference on Intelligent Robots and Systems, pp. 3181–3186 (2005)
27. Choi, J., Oh, S., Horowitz, R.: Cooperatively learning mobile agents for gradient climbing. In: the Proceedings of 2007 46th IEEE Conference on Decision and Control, pp. 3139–3144 (2007)
28. Chopra, N., Spong, M.: Passivity-based control of multi-agent systems. In: S. Kawamura, M. Svinin (eds.) Advances in Robot Control, From Everyday Physics to Human-Like Movements, pp. 107–134. Springer-Verlag, Berlin (2006)
29. Chopra, N., Spong, M.: Output synchronization on strongly connected graphs. In: MTNS (2008)
30. Chow, J. (ed.): Time Scale Modelling of Dynamic Networks with Applications to Power Systems, Lecture Notes in Control and Information Sciences, vol. 46. Springer Verlag, New York (1982)
31. Chung, S.J., Ahsun, U., Slotine, J.: Application of synchronization to formation flying spacecraft:lagrangian approach. Journal of Guidance, Control and Dynamics **32**, 512–526 (2009)
32. Chung, S.J., Slotine, J.: Cooperative robot control and concurrent synchronization of lagrangian systems. IEEE Transations on Robotics **25**, 686–700 (2009)
33. Cortés, J.: Distributed Kriged Kalman filter for spatial estimation. IEEE Transactions on Automatic Control **54**(12), 2816–2827 (2009)
34. Cortés, J., Martínez, S., Karataş, T., Bullo, F.: Coverage control for mobile sensing networks. IEEE Transactions on Robotics and Automation **20**(2), 243–255 (2004)
35. Dennis, J.E., Schnabel, R.B.: Numerical Methods for Unconstrained Optimization and Nonlinear Equations. SIAM, Philadelphia (1996)
36. Desai, J., Ostrowski, J., Kumar, V.: Modeling and control of formations of nonholonomic mobile robots. IEEE Transactions on Robotics and Automation **17**(6), 905–908 (2001)
37. Dimarogonas, D.V., Kyriakopoulos, K.J.: On the rendezvous problem for multiple nonholonomic agents. Automatic Control, IEEE Transactions on **52**(5), 916–922 (2007)
38. Dimarogonas, D.V., Tsiotras, P., Kyriakopoulos, K.J.: Laplacian cooperative attitude control of multiple rigid bodies. In: IEEE International Symposium on Intelligent Control., pp. 3064–3069 (2006)
39. Do, K.D.: Formation tracking control of unicycle-type mobile robots with limited sensing ranges. IEEE Transactions on Control System Technology **16**(3), 527–538 (2008)
40. Do, K.D., Jiang, Z.P., Pan, J.: Robust adaptive path following of underactuated ships. In: Proceedings of the 41st IEEE Conference on Decision and Control, pp. 3243–3248. Las Vegas, NV, USA (2002)
41. Dong, W., Farrell, J.A.: Cooperative control of multiple nonholonomic mobile agents. IEEE Transactions On Automatic Control **53**(6), 1434–1448 (2008)
42. Dong, W., Farrell, J.A.: Decentralized cooperative control of multiple nonholonomic dynamic systems with uncertainty. Automatica **45**(3), 706–710 (2009)
43. Dörfler, F.: Geometric analysis of the formation control problem for autonomous robots. Diploma thesis, University of Toronto (2008)
44. Dunbar, W.B., Murray, R.M.: Distributed receding horizon control for multi-vehicle formation stabilization. Automatica **42**(4), 549–558 (2006)
45. Encarnação, P., Pascoal, A.: Combined trajectory tracking and path-following: An application to the coordinated control of autonomous marine craft. In: Proceedings of the 40th IEEE Conference on Decision and Control. Orlando, FL, USA (964-969)
46. Eren, T., Belhumeur, P., Anderson, B., Morse, A.: A framework for maintaining formations based on rigidity. In: Proceedings of IFAC World Congress (2002)

47. Fax, J.A., Murray, R.M.: Information flow and cooperative control of vehicle formations. IEEE Transactions on Automatic Control **49**(9) (2004)
48. Fierro, R., Das, A.E., Kumar, V., Ostrowski, J.P.: Hybrid control of formations of robots. In: Proceedings of the 2001 IEEE International Conference on Robotics and Automation, pp. 157–162. Seoul, Korea (2001)
49. Fossen, T.I.: Marine Control Systems: Guidance, Navigation and Control of Ships, Rigs and Underwater Vehicles. Marine Cybernetics, Trondheim, Norway
50. Gazi, V.: Formation control of a multi-agent system using non-linear servomechanism. International Journal of Control **78**(8), 554–565 (2005)
51. Gazi, V., Passino, K.M.: Stability analysis of social foraging swarms. IEEE Transactions on Systems, Man and Cybernetics - Part B: Cybernetics **34**(1) (2004)
52. Hale, J.K.: Oscillations in Nonlinear Systems. McGraw-Hill Book Company, Inc. (1963)
53. Hale, J.K.: Ordinary Differential Equations, second edn. Robert E. Krieger Publishing Company, INC. (1980)
54. Hauser, J., Hindman, R.: Maneuver regulation from trajectory tracking: feedback linearizable systems. In: Proceedings of IFAC Symposium on Nonlinear Control Systems Design, pp. 595–600. Lake Tahoe, CA (1995)
55. Hendrickx, J.M., Anderson, B.D.O., Delvenne, J.C., Blondel, V.D.: Directed graphs for the analysis of rigidity and persistence in autonomous agent systems. International Journal of Robust Nonlinear Control **17**, 960–981 (2007)
56. Hespanha, J.P., Morse, A.S.: Stability of switched systems with average dwell-time. Tech. rep., Electrical Eng. Dept., Univ. of Southern California (1999). Available at `http://www.ece.ucsb.edu/~hespanha/techrep.html`
57. Hirata, Y., Kume, Y., Wang, Z., Kosuge, K.: Decentralized control of multiple mobile manipulators handling a single object in coordination. In: Proceedings of IEEE International Conference on Intelligent Robots and Systems, pp. 2758–2763 (2002)
58. Igarashi, Y., Hatanaka, T., Fujita, M., Spong, M.W.: Passivity-basd attitude synchronization in $SE(3)$. IEEE Transactions on Control Systems Technology **17**(5) (2009)
59. Ihle, I.A.F., Arcak, M., Fossen, T.I.: Passivity-based designs for synchronized path-following. Automatica **43**(9), 1508 – 1518 (2007)
60. Ihle, I.A.F., Skjetne, R., Fossen, T.I.: Nonlinear formation control of marine craft with experimental results. In: Proceedings of the IEEE Conference on Decision and Control, pp. 680–685. Atlantis, Paradise Island, The Bahamas (2004)
61. Ioannou, P., Tao, G.: Frequency domain conditions for strictly positive real functions. IEEE Transaction on Automatic Control **32**(10), 53–54 (1987)
62. Ioannou, P.A., Sun, J.: Robust Adaptive Control. Prentice Hall (1996)
63. Jadbabaie, A., Lin, J., Morse, A.S.: Coordination of groups of mobile autonomous agents using nearest neighbor rules. IEEE Transactions on Automatic Control **48**(6), 988–1001 (2003)
64. Jayawardhana, B.: Tracking and disturbance rejection for passive nonlinear systems. In: Proceedings of the 44th IEEE Conference on Decision and Control, and the European Control Conference, pp. 3333–3338 (2005)
65. Jayawardhana, B., Ortega, R., García-Canseco, E., Castaños, F.: Passivity of nonlinear incremental systems: Application to pi stabilization of nonlinear rlc circuits. Systems and Control Letters **56**, 618–622 (2007)
66. Kang, W., Yeh, H.H.: Co-ordinated attitude control of multi-satellite systems. International Journal of Robust and Nonlinear Control **12**, 185–205 (2002)
67. Kelley, C.T.: Iterative Methods for Linear and Nonlinear Equations. SIAM, Philadelphia (1995)
68. Kelly, R.: Comments on "Adaptive PD controller for robot manipulators". IEEE Transactions on Robotics and Automation **9**(1), 117–119 (1993)
69. Khalil, H.: Nonlinear Systems, 3rd edn. Prentice Hall, Upper Saddle River, NJ (2002)
70. Khosla, P., Kanade, T.: Parameter identification of robot dynamics. In: Proceedings of the 24th IEEE Conference on Decision and Control. Fort Lauderdale, FL (1985)

71. Koditschek, D.: Natural motion of robot arms. In: IEEE Conference on Decision and Control. Las Vegas, NV (1984)
72. Krick, L., Broucke, M., Francis, B.: Stabilization of infinitesimally rigid formations of multi-robot networks. International Journal of Control **82**, 423–439 (2009)
73. Labeyrie, A., Lipson, S. G., Nisenson, P.: An Introduction to Optical Stellar Interferometry. Cambridge University Press, New York (2006)
74. Lau, K., Lichten, S., Young, L., Hanes, B.: An innovative deep space applications of gps technology for formation flying spacecraft. In: AIAA Guidance, Navigation and Control Conference. San Diego, CA (1996)
75. Lawton, J.R., Beard, R.W.: Synchronized multiple spacecraft rotations. Automatica **38**, 1359–1364 (2002)
76. Lee, D., Spong, M.W.: Stable flocking of multiple inertial agents on balanced graphs. IEEE Transactions on Automatic Control **52**(8), 1469–1475 (2007)
77. Leonard, N., Paley, D., Lekien, F., Sepulchre, R., Fratantoni, D., Davis, R.: Collective motion, sensor networks, and ocean sampling. Proceedings of the IEEE **95**(1), 48–74 (2007)
78. Li, W., Cassandras, C.G.: Distributed cooperative coverage control of sensor networks. In: the Proceedings of the 44th IEEE Conference on Decision and Control, pp. 2542–2547 (2005)
79. Li, Z., Duan, Z., Chen, G., Huang, L.: Consensus of multiagent systems and synchronization of complex networks: a unified viewpoint. IEEE Transactions on Circuits and Systems Part I **57**(1), 213–224 (2010)
80. Liberzon, D.: Switching in Systems and Control. Birkhauser, Boston, MA (2003)
81. Liberzon, D., Morse, A.: Basic problems in stability and design of switched system. IEEE Control Systems Magazine **19**, 59–70 (1999)
82. Lin, Y., Sontag, E., Wang, Y.: A smooth converse Lyapunov theorem for robust stability. SIAM Journal of Control and Optimization **34**, 124–160 (1996)
83. Lin, Z., Francis, B., Maggiore, M.: Necessary and sufficient graphical conditions for formation control of unicycles. IEEE Trans. on Automatic Control **50**(1), 121–127 (2005)
84. Lizarralde, F., Wen, J.T.: Attitude control without angular velocity measurement: A passivity approach. IEEE International Conference on Robotics and Automation **41**(3) (1996)
85. Loría, A., Panteley, E., Popović, D., Teel, A.R.: A nested Matrosov theorem and persistency of excitation for unifom convergence in stable nonautonomous systems. IEEE Transactions on Automatic Control **50**(2) (2005)
86. Lynch, K.M., Schwartz, I.B., Yang, P., Freeman, R.A.: Decentralized environment modeling by mobile sensor networks. IEEE Transactions on Robotics **24**(3), 710–724 (2008)
87. Marshall, J., Broucke, M., Francis, B.: Formation of vehicles in cyclic pursuit. IEEE Transactions on Automatic Control **49**(11), 1963–1974 (2004)
88. Mastellone, S., Stipanović, D.M., Graunke, C.R., Intlekofer, K.A., Spong, M.W.: Formation control and collision avoidance for multi-agent non-holonomic systems: Theory and experiments. Int. J. Rob. Res. **27**, 107–126 (2008)
89. Montemayor, G., Wen, J.T.: Decentralized collaborative load transport by multiple robots. In: Proceedings of 2005 IEEE Robotics and Automation Conference, Barcelona, Spain (2005)
90. Moreau, L.: Stability of multiagent systems with time-dependent communication links. IEEE Transactions on Automatic Control **50**(2), 169–182 (2005)
91. Morse, A.S.: Supervisory control of families of linear set-point controllers, part i: Exact matching. IEEE Transactions on Automatic Control **41**(10), 1413–1431 (1996)
92. Munawar, K., Uchiyama, M.: Experimental verification of distributed event-based control of multiple unifunctional manipulators. In: Proceedings of IEEE International Conference on Robotics and Automation, pp. 1213–1218 (1999)
93. Murray, R.M.: Recent Research in Cooperative Control of Multivehicle Systems. Journal of Dynamic Systems, Measurement, and Control **129**(5), 571–583 (2007)
94. Murray, R.M., Li, Z., Sastry, S.S.: A Mathematical Introduction to Robotic Manipulation. CRC Press (1994)

95. Nair, S., Leonard, N.E.: Stabilization of a coordinated network of rotating rigid bodies. In: Proceedings of 43rd IEEE Conference on Decision and Control. Atlantic, Paradise Island, Bahamas, (2004)
96. Nayfeh, A.H., Mook, D.T.: Nonlinear Oscillations. John Wiley, New York (1979)
97. Nelson, T., Freeman, R.: Decentralized $\mathscr{H}_\infty$ filtering in a multi-agent system. In: Proceedings of the 2009 American Control Conference, pp. 5755–5760. St. Louis, Missouri (2009)
98. Ögren, P., Fiorelli, E., Leonard, N.E.: Cooperative control of mobile sensor networks: Adaptive gradient climbing in a distributed network. IEEE Transactions on Automatic Control **49**(8), 1292–1302 (2004)
99. Olfati-Saber, R.: Distributed Kalman filter with embedded consensus filters. In: The 44th IEEE Conference on Decision and Control and 2005 European Control Conference, pp. 8179–8184 (2005)
100. Olfati-Saber, R.: Flocking for multi-agent dynamic systems: Algorithms and theory. IEEE Transactions on Automatic Control **51**(3), 401–420 (2006)
101. Olfati-Saber, R., Murray, R.M.: Graph rigidity and distributed formation stabilization of multi-vehicle systems. In: Proceedings of the IEEE Int. Conference on Decision and Control, pp. 2965–2971 (2002)
102. Olfati-Saber, R., Murray, R.M.: Consensus protocols for networks of dynamic agents. In: Proceedings of 2003 American Control Conference, pp. 951–956 (2003)
103. Olfati-Saber, R., Murray, R.M.: Consensus problems in networks of agents with switching topology and time-delays. IEEE Transactions on Automatic Control **49**(9) (2004)
104. Olfati-Saber, R., Shamma, J.S.: Consensus filters for sensor networks and distributed sensor fusion. In: Proceedings of the 44th IEEE Conference on Decision and Control and European Control Conference, pp. 6698–6703. Seville, Spain (2005)
105. Ortega, R., Loría, A., Nicklasson, P.J., Sira-Ramírez, H.: Passivity-based control of Euler–Lagrange systems, vol. Communications and Control Engineering. Springer-Verlag, Berlin (1998)
106. Pereira, G.A.S., Kumar, V., Campos, M.F.M.: Decentralized algorithms for multirobot manipulation via caging. International Journal of Robotics Research (2004)
107. Reif, J.H., Wang, H.: Social potential fields: A distributed behavioral control for autonomous robots. Robotics and Autonomous Systems **27**, 171–194 (1999)
108. Ren, W.: Synchronized multiple spacecraft rotations: A revisit in the context of consensus building. In: Proceedings of the 2007 American Control Conference, pp. 3174–3179 (2007)
109. Ren, W., Beard, R.: Distributed Consensus in Multi-vehicle Cooperative Control: Theory and Applications. Communications and Control Engineering. Springer-Verlag, London (2008)
110. Ren, W., Beard, R.W., Atkins, E.M.: A survey of consensus problems in multi-agent coordination. In: Proceedings of 2005 American Control Conference, pp. 1859–1864 (2005)
111. Ren, W., Moore, K.L., Chen, Y.: High-order and model reference consensus algorithms in cooperative control of multi-vehicle systems. ASME Journal of Dynamic Systems, Measurement, and Control **129**(5), 678–688 (2007)
112. Reynolds, C.: Flocks, herds, and schools: A distributed behavioral model. Computer Graphics **21**, 25–34 (1987)
113. Rimon, E., Koditschek, D.E.: Exact robot navigation using artificial potential functions. IEEE Transactions on Robotics and Automation **8**(5), 501–518 (1992)
114. Samson, C.: Path following and time-varying feedback stabilization of a wheeled mobile robot. In: Proceedings of the ICARCV. Singapore (1992)
115. Sarlette, A., Sepulchre, R., Leonard, N.E.: Autonomous rigid body attitude synchronization. In: Proceedings of the 46th IEEE Conference on Decision and Control. New Orleans, LA (2007)
116. van der Schaft, A.: $\mathscr{L}_2$-Gain and Passivity Techniques in Nonlinear Control, second edn. Springer-Verlag (1999)
117. Scharf, D.P., Hadaegh, F.Y., Ploen, S.R.: A survery of spacecraft formation flying guidance and control (part I): Guidance. In: Proceedings of the American Control Conference, pp. 1733–1739. Denver, Colorado (2003)

118. Scharf, D.P., Hadaegh, F.Y., Ploen, S.R.: A survey of spacecraft formation flying guidance and control (part II): Control. In: Proceedings of the American Control Conference, pp. 2976–2985 (2004)
119. Sepulchre, R., Janković, M., Kokotoivć, P.: Constructive Nonlinear Control. New York: Springer-Verlag (1997)
120. Skjetne, R., Fossen, T.I., Kokotović, P.V.: Robust output maneuvering for a class of nonlinear systems. Automatica **40**(3), 373–383 (2004)
121. Slotine, J., Li, W.: On the adaptive control of robot manipulators. International Journal of Robotics Research **6**(3), 147–157 (1987)
122. Slotine, J.J.E., Wang, W.: A study of synchronization and group coordination using partial contraction theory. In: V. Kumar, N. Leonard, A.Morse (eds.) Cooperative Control, vol. 309, pp. 207–228. Springer-Verlag (2005)
123. Smith, R.S., Hadaegh, F.Y.: Control of deep-space formation-flying spacecraft; relative sensing and switched information. Journal of Guidance, Control of Dynamics **28**(1), 106–114 (2005)
124. Smith, R.S., Hadaegh, F.Y.: Closed-loop dynamics of cooperative vehicle formations with parallel estimators and communication. IEEE Transactions on Automatic Control **52**(8) (2007)
125. Sontag, E.D.: Smooth stabilization implies coprime factorization. IEEE Transactions on Automatic Control **34**(4), 435–443 (1989)
126. Sontag, E.D.: Further facts about input to state stabilization. IEEE Transactions on Automatic Control **35**, 473–476 (1990)
127. Sorensen, N., Ren, W.: A unified formation control scheme with a single or multiple leaders. In: Proceedings of the 2007 American Control Conference. New York City, New York (2007)
128. Sugar, T.G., Kumar, V.: Control of cooperating mobile manipulators. IEEE Transactions on Robotics and Automation **18**(1), 94–103 (2002)
129. Sun, D., Mills, J.K.: Manipulating rigid payloads with multiple robots using compliant grippers. IEEE/ASME Transactions on Mechatronics **7**(1), 23–34 (2002)
130. Tang, C.P., Bhatt, R.M., Abou-Samah, M., Krovi, V.: Screw-theoretic analysis framework for cooperative payload transport by mobile manipulator collectives. IEEE/ASME Transactions on Mechatronics **11**(2), 169–178 (2006)
131. Tanner, H., Jadbabaie, A., Pappas, G.: Flocking in fixed and switching networks. IEEE Transactions on Automatic Control **52**(5), 863–868 (2007)
132. Tanner, H., Kumar, A.: Towards decentralization of multi-robot navigation functions. In: Proceedings of the 2005 IEEE International Conference on Robotics and Automation, pp. 4132 – 4137 (2005)
133. Tanner, H.G., Kumar, A.: Formation stabilization of multiple agents using decentralized navigation functions. In: Proceedings of Robotics: Science and Systems. Cambridge, MA, USA (2005)
134. Tanner, H.G., Kyriakopoulos, K.J., Krikelis, N.I.: Modeling of multiple manipulators handling a common deformable object. Journal of Robotic Systems **15**(11), 599–623 (1998)
135. Tanner, H.G., Pappas, G.J., Kumar, V.: Leader-to-formation stability. IEEE Transactions on Robotics and Automation **20**(3), 443–455 (2004)
136. Tao, G.: Adaptive Control Design and Analysis. John Wiley & Sons (2003)
137. Teel, A., Praly, L.: A smooth Lyapunov function from a class-KL estimate involving two positive semidefinite functions. ESAIM: Control, Optimisation and Calculus of Variations **5**, 313–367 (2000)
138. Tomei, P.: Adaptive PD controller for robot manipulators. IEEE Transactions On Robotics And Automation **7**(4), 565–570 (1991)
139. Tsiotras, P.: Further passivity results for the attitude control problem. IEEE Transactions on Automatic Control **43**(11) (1998)
140. Tsitsiklis, J.N.: Problems in decentralized decision making and computation. PhD Thesis, Department of EECS, MIT (1984)
141. VanDyke, M.C.: Decentralized coordinated attitude control of a formation of spacecraft. Master thesis, Virginia Polytechnic Institute and State University, Blacksburg, VA (2004)

142. Wen, J.T.: Time domain and frequency domain conditions for strict positive realness. IEEE Transaction on Automatic Control **33**(10), 988–992 (1988)
143. Wen, J.T.: A unified perspective on robot control: The energy lyapunov function approach. International Journal on Adaptive Control and Signal Processing **4**(6), 487–500 (1990)
144. Wen, J.T., Kreutz-Delgado, K.: The attitude control problem. IEEE Transactions on Automatic Control **36**(10), 1148–1162 (1991)
145. Wen, J.T., Kreutz-Delgado, K.: Motion and force control of multiple robotic manipulators. Automatica **28**(4), 729–743 (1992)
146. Wittenburg, J.: Dynamics of Systems of Rigid Bodies. Stuttgard, West Germany: B.G. Teubner (1977)
147. Wu, C.: Synchronization in arrays of coupled nonlinear systems: Passivity, circle criterion, and observer design. IEEE Transactions on Circuits and Systems - I: Fundamental Theory and Applications **48**(10), 1257–1261 (2001)
148. Xiao, L., Boyd, S.: Fast linear iterations for distributed averaging. Systems and Control Letters **53**, 65–78 (2004)
149. Xie, W.C.: Dynamic Stability of Structures. Cambridge University Press (2006)
150. Yang, P., Freeman, R.A., Lynch, K.M.: Mutli-agent coordination by decentralized estimation and control. IEEE Transactions on Automatic Control **53**(11), 2480–2496 (2008)
151. Zhang, C., Arnold, D., Ghods, N., Siranosian, A., Krstić, M.: Source seeking with non-holonomic unicycle without position measurement and with tuning of forward velocity. Systems & Control Letters (56) (2007)
152. Zhang, F.: Geometric Cooperative Control of Particle Formations. IEEE Transactions on Automatic Control **55**(3), 800–804 (2010)
153. Zhao, F., Guibas, L.: Wireless Sensor Networks: An Information Processing Approach, first edn. The Morgan Kaufmann Series in Networking, San Francisco, CA (2004)
154. Zheng, Y., Luh, Y.: Control of two coordinated robots in motion. In: Proceedings of the 25th IEEE Conference on Decision and Control, pp. 334–337. Ft. Lauderdale, FL (1985)

FSC
www.fsc.org
MIX
Papier | Fördert
gute Waldnutzung
FSC® C083411

Kolibri 360 GmbH
Ferdinand-Jühlke-Straße 7
99095 Erfurt, Deutschland
produktsicherheit@kolibri360.de